A C

"In the age of digit... ...dical specialty needs an author like Praveen Suthrum and a book like this to present a clear vision of how that particular field can embrace advanced technologies while focusing on the doctor-patient relationship. Such future-oriented thinking is crucial for making professionals in that field more prepared and resilient."

- *Dr. Bertalan Mesko, The Medical Futurist, director of The Medical Futurist Institute*

"**FABULOUS! This book will definitely be on everyone's reading list. *The Shift* will open your mind to the exponential forces shaping our rapidly evolving healthcare landscape.** It will wake you up to the pace of innovation charging at us faster than ever. A must-read for anyone interested in the expanding practice and business of GI wellness!"

- *Dan Neumann, MD, president and chief strategy officer, Capital Digestive Care*

"**Compelling, bold, and most importantly, REAL.** Praveen has cultivated a community of fantastic innovators over many years, drawing out stories, battle scars, and learnings that span the entirety of gastrointestinal innovation. This book is a must-read for the aspiring innovator, the gastroenterologist, and anyone who wants to make a difference in the future of healthcare."

- *Dr. Jonathan Ng, founder and CEO, Iterative Health*

"**This book is not just about watching the future unfold; it's about shaping it.** Praveen's vision for a future where the exponential advancements in AI and digital health revolutionize gastroenterology and beyond is both inspiring and practical. As we stand at the cusp of a new era, *The Shift* offers a vital roadmap for anyone committed to leading the next wave of healthcare innovation."

- Daniel Kraft, MD, founder of NextMed Health and Digital.Health

"Our universal human response is to fear change – especially in an industry that ethically cannot 'move fast and break things,' as innovation culture thrives in other sectors. Yet here we stand, facing the necessity of change and the opportunities presented by digital innovation. The path forward requires asking hard and critical questions and considering diverse ideas from innovators on the frontline in gastroenterology. Praveen does both so well, and **this book is essential reading as we aspire to the Quadruple Aim of high patient and clinician satisfaction, great clinical outcomes, and making healthcare accessible and affordable.**"

- Sam Holliday, cofounder and CEO, Oshi Health

"*The Shift* **is a groundbreaking exploration of how digital innovation is revolutionizing the field of gastroenterology** and the broader healthcare landscape. This book is a must-read for anyone invested in the future of medicine. It expertly blends cutting-edge research with real-world applications, providing a comprehensive guide to the technologies reshaping patient care. An enlightening and inspiring read highlighting the immense potential of digital advancements in transforming healthcare for the better."

- Erica Barnell, MD, PhD, cofounder, chief medical and science officer, Geneoscopy

"In the realm of gastroenterology, **we find ourselves at a juncture reminiscent of the early days of space exploration**, where rapid advancements defined new territories. We stand at a critical inflection point, where inaction risks the obsolescence of traditional practices. ***The Shift* masterfully uses storytelling to illuminate disruptive innovations and the visionaries behind them**, providing a framework that stirs excitement and propels us toward transformative progress."

- Aja McCutchen, MD, board member, United Digestive

"**Praveen's willingness to make bold, and sometimes controversial, projections about the current and future state of GI and healthcare is of huge value to the field.** His ideas regularly challenge and inspire me to consider new perspectives. Perhaps more importantly, though, Praveen has admirably stepped up to put a microphone in front of so many other unique voices across the industry. In doing so, he created a powerful library of viewpoints — these interviews are simply a must-read for anyone interested in GI innovation."

- Matt Schwartz, cofounder and CEO, Virgo

"Hot on the heels of his previous success with the publication of *Scope Forward*, Praveen has put together another truly compelling resource that highlights the rapid progress and huge opportunity for digital innovation in the field of gastroenterology. He has once again brought together a hugely impressive list of successful entrepreneurs in this space. **Warning---he makes it very clear in this new book that there is no putting the lid back on this one and that digital transformation in gastroenterology and healthcare, in general, is here to stay.** Enjoy!"

- Dr. Michael Byrne, chairman, founder, and CMO, Satisfai Health

"**This book is a bold and honest conversation, and Praveen is not scared to ask questions that the industry shies away from or that make many of us uneasy.** Disruption in GI and the medical industry is here, will stay, and will continue to be transformative to all our careers, whether we are physicians, entrepreneurs, or scientists in the healthcare field. Praveen gives us a very relevant snapshot of the industry shifts, along with words of wisdom from the collective intellect that is the very momentum of this change. What truly inspires is Praveen's deeply personal and honest recounting of his struggles and triumphs, as well as the arduous journey he undertook to become the person he is today. Truly a blockbuster. Dive in!"

- *Treta Purohit, MD, gastroenterologist, digital gastroenterology consultant*

"Change is hard, and the status quo is always the most formidable competitor. *The Shift* is a must-read for anyone who is interested in looking around the corner and taking on the challenge of moving gastroenterology forward. **Praveen's insights are not only deeply informative but also inspiring**, pushing us to embrace change and rethink what's possible in the future."

- *Russ Arjal, MD, cofounder and CMO, WovenX Health*

THE SHIFT

HOW DIGITAL INNOVATION IS IRREVERSIBLY TRANSFORMING GASTROENTEROLOGY AND HEALTHCARE

PRAVEEN SUTHRUM

AUTHOR OF SCOPE FORWARD AND
PRIVATE EQUITY IN GASTROENTEROLOGY

THE SHIFT © 2024 by Praveen Suthrum. All rights reserved.

Printed in the United States of America

Published by Igniting Souls
PO Box 43, Powell, OH 43065
IgnitingSouls.com

This book contains material protected under international and federal copyright laws and treaties. Any unauthorized reprint or use of this material is prohibited. No part of this book may be reproduced or transmitted in any form or by any means, electronic or mechanical, including photocopying, recording, or by any information storage and retrieval system, without express written permission from the author.

LCCN: 2024919300
Paperback ISBN: 978-1-63680-368-5
Hardcover ISBN: 978-1-63680-369-2
e-book ISBN: 978-1-63680-370-8

Available in paperback, hardcover, e-book, and audiobook.

Any Internet addresses (websites, blogs, etc.) and telephone numbers printed in this book are offered as a resource. They are not intended in any way to be or imply an endorsement by Igniting Souls, nor does Igniting Souls vouch for the content of these sites and numbers for the life of this book.

Some names and identifying details may have been changed to protect the privacy of individuals.

For Suguna and Murthy, my wonderful parents.
Thank you for being you.

See, I wrote this without *effeffeus!*

CONTENTS

WE'RE LIVING IN THE ERA OF THE SECOND DERIVATIVE - FOREWORD BY DR. LAWRENCE KOSINSKI 1

INTRODUCTION 7

SECTION: AN IRREVERSIBLE SHIFT

1. THE DIGITAL SHIFT: FROM TRADITIONAL TO TRANSFORMATIVE 17
   COVID-19 LAID THE FOUNDATION FOR DIGITAL HEALTH 24
   A SHIFT FROM SCARCITY TO ABUNDANCE 30
   THIS IS A SHACKLETON MOMENT 35

2. THE BUSINESS SHIFT: FROM LINEAR TO EXPONENTIAL MODELS .. 39
   UNDERSTANDING EXPONENTIAL MODELS 45
   WHERE DOES THE SHIFT TO AN EXPONENTIAL MODEL BEGIN? 54

3. THE MINDSET SHIFT: FROM BURNOUT TO INNOVATION 61
   THE SHIFT TO DIGITAL MAGNIFIES EVERYTHING, INCLUDING THE FEELING OF DISILLUSIONMENT 63
   STORIES OF INNOVATION AND RESILIENCE IN GASTROENTEROLOGY 65
   TO SPOT FUTURE OPPORTUNITIES, ZOOM OUT 74

4. THE DECISIVE SHIFT: FROM WATCHING TO SHAPING THE FUTURE .. 79
   THIS SHIFT ISN'T ABOUT GASTROENTEROLOGY OR HEALTHCARE ALONE 82
   THE IMPENDING AI REVOLUTION IS MUCH CLOSER THAN YOU THINK 84
   A DAY IN THE LIFE OF A GASTROENTEROLOGIST IN 2034 88
   THE SHIFT WITHIN 96

SECTION: VOICES I - DIAGNOSTICS AND DEVICES

DIAGNOSING DISEASE THROUGH BREATH:
AONGHUS SHORTT OF FOODMARBLE 101
TRACKING GUT'S ELECTRICAL ACTIVITY: DR. GREG O'GRADY OF ALIMETRY.... 110
BUILDING SMART TOILETS: DR. SONIA GREGO OF COPRATA 120
BREAKING BARRIERS IN CANCER SCREENING THROUGH NON-INVASIVE
DIAGNOSTICS: DR. ERICA BARNELL OF GENEOSCOPY 128
DEMOCRATIZING ENDOSCOPIC ULTRASOUND TECHNOLOGY:
DR. STEPHEN STEINBERG OF ENDOSOUND 141

SECTION: VOICES II - AI AND ROBOTICS

HEALTHCARE EQUITY THROUGH AI IN GI:
DR. JONATHAN NG OF ITERATIVE HEALTH 155
ADDING A DIGITAL LAYER TO ENDOSCOPY: MATT SCHWARTZ OF VIRGO 165
THE ELEPHANT IN THE ROOM IS REIMBURSEMENT FOR AI ADOPTION:
DR. MICHAEL BYRNE OF SATISFAI HEALTH 177
BRINGING AI TO THE ENDOSCOPY ROOM THROUGH GI GENIUS:
DR. ANDREA CHERUBINI OF COSMO IMD 189
FROM SELF-DRIVING CARS TO SELF-DRIVING ENDOSCOPES:
SAURABH JEJURIKAR OF ENDOVISION AI 199
DON'T FORGET THAT THE ROBOT IS PART OF THE TEAM:
DR. SANKET CHAUHAN OF SURGICAL AUTOMATIONS 205
DRONES THAT SWIM IN YOUR GUT: TORREY SMITH OF ENDIATX (PILLBOT) ... 214

SECTION: VOICES III - DATA, APPS, AND GAMIFICATION

TRANSFORMING GI DATA INTO AN ANCILLARY: OMER DROR OF LYNX.MD .. 225
FROM MACHINE LEARNING AT UBER TO A STOOL APP:
ASAF KRAUS OF DIETA HEALTH 237
UNLOCKING THE VALUE OF DATA FROM GI SURGERY CENTERS:
DUNSTON ALMEIDA OF TRIVALENCE 250
GAMIFYING GI DATA ANNOTATION FOR AI:
ERIK DUHAIME OF CENTAUR LABS 256

SECTION: VOICES IV - NEW CARE MODELS

FROM GI PATIENT TO DIGITAL HEALTH CEO:
SAM JACTEL OF AYBLE HEALTH 269
HYBRID GI CARE IS HERE TO STAY:
SAM HOLLIDAY AND DR. TRETA PUROHIT OF OSHI HEALTH 282
WHAT IF WE DIDN'T DO AS MANY COLONOSCOPIES?
DR. MICHAEL OWENS OF PEARL HEALTH PARTNERS. 294
BRIDGING THE BRAIN-GUT CONNECTION:
DR. MEGAN RIEHL AND DR. MADISON SIMONS 304
VIRTUAL REALITY THERAPEUTICS:
DR. BRENNAN SPIEGEL, AUTHOR OF VRX. 312

SECTION: VOICES V - ENTREPRENEURSHIP AND CAREERS

LEAVING PRIVATE PRACTICE TO PURSUE LIFE: DR. FEHMIDA CHIPTY 325
BEYOND LINEAR HEALTHCARE MODELS:
DR. DAN NEUMANN OF TRILLIUM HEALTH. 335
A DIGITAL STARTUP FOR WOMEN'S GUT ISSUES:
DR. AJA MCCUTCHEN OF OLVI HEALTH AND UNITED DIGESTIVE. 345
FROM THE ESOPHAGUS TO THE COLON, GI HAS VAST SCOPE
FOR INNOVATION: DR. BARA EL KURDI 354
BUILDING AN END-TO-END VIRTUAL GI STARTUP:
DR. RUSS ARJAL OF WOVENX HEALTH. 366

SECTION: VOICES VI - PERSPECTIVES

"GI FACTORIES" ARE MORE VULNERABLE NOW:
DR. LAWRENCE KOSINSKI OF SONARMD. 383
THE FUTURE OF GI REMAINS BRIGHT:
DR. SCOTT KETOVER OF MNGI DIGESTIVE HEALTH 394
MOVING ON FROM GERM THEORY TO PRECISION MEDICINE:
PROF. DAVID WHITCOMB OF ARIEL PRECISION MEDICINE. 404
WE NEED MORE HEALTH PROFESSIONALS ON SOCIAL MEDIA:
DR. AUSTIN CHIANG OF MEDTRONIC ENDOSCOPY. 416

ACKNOWLEDGMENTS 425
NOTES 427
ABOUT THE AUTHOR 447

WE'RE LIVING IN THE ERA OF THE SECOND DERIVATIVE - FOREWORD BY DR. LAWRENCE KOSINSKI

In calculus, the second derivative measures the rate at which the rate of change itself is changing. This concept is not merely academic but profoundly illustrative of the accelerating pace of innovation and adaptation required in today's healthcare environment. We are now living in an era of an increasing second derivative.

To complicate matters further, there are multiple strategic forces impacting us today, each of which is in its own second derivative of change. This sets the stage for logarithmic change. The combined effect of these forces on us is resulting in stress on the system and burnout for providers.

What are these forces? I see six major categories.

STRUCTURAL CHANGES IN HEALTHCARE LEGISLATION

Most of the changes imposed on our healthcare system are a result of federal legislation. The Recovery Act in 2009 gave us Meaningful Use, which promoted the conversion of paper records to digital health records. Although this was meant to help coordinate care, what we have today are fragmented systems that often hinder communication between providers.

Following this, the Affordable Care Act (ACA) was passed, which included the Center for Medicare and Medicaid Innovation (CMMI), to test new payment and delivery models that would lower spending while maintaining or improving care. $10 billion was allocated for the years 2011 to 2019, followed by another $10 billion for the subsequent decade. Over the last decade, CMMI has spearheaded over 50 pilot projects, including the Medicare Shared Savings Program (MSSP), reaching nearly 28 million patients and over 528,000 healthcare providers and plans. These models provide important lessons on transitioning the US health system to a value-based care model.

The Medicare Access and CHIP Reauthorization Act of 2015 (MACRA) further aimed to increase accountability in healthcare, establishing the Physician-Focused Payment Model Technical Advisory Committee (PTAC) to evaluate new payment models. Despite initial hopes, to date, CMS has not implemented specialty-focused, episode-focused, or disease-specific alternative payment models, continuing to prioritize population-based total cost of care models.

CONSOLIDATION

Health plans began consolidating in the 1990s, significantly enhancing their market power and influence. The ACA also facilitated accelerated consolidation in healthcare, significantly increasing the influence of large health systems. Today, approximately five major health plan platforms dominate the commercial healthcare space, which, in turn, increases physician employment in these systems. In parallel, historically low interest rates over the last decade led to a wave of private-equity-led acquisitions in the gastroenterology (GI) private practice space. There have been no initial public offerings (IPOs) with these investments to date.

PAYMENT MODEL CHANGES

Since the introduction of the Resource-Based Relative Value Scale, physician reimbursement has been subject to annual revisions. The structural changes created by the ACA and MACRA paved the way for a transition from fee-for-service to value-based care. Despite these intentions, the practical outcome has been a 23 percent decline in professional reimbursement rates since 2000. Risk of healthcare financing is also shifting from Medicare to Medicare Advantage and Medicare Shared Savings Programs (ACOs). Similarly, in the commercial sector, health plans have transferred risk to ACOs and self-funded employers, fostering a new category of healthcare providers: the "payvider." These new "payviders," who provide care and assume financial risk, are the model for future accountable care. This shift necessitates the development of nested risk-bearing solutions even within specialties such as gastroenterology.

DISRUPTIVE TECHNOLOGIES

The initial promise of digital health records was transformative: to revolutionize healthcare delivery, improve patient outcomes, and streamline costs. Unfortunately, rather than empowering physicians, these systems have relegated them to the role of data entry technicians. The silos of data that do not communicate with each other have created a nightmare of redundant data entry for providers.

The COVID-19 pandemic catalyzed the adoption of telehealth, enabling the possibility of remote care. This shift exposed the limitations of practice-based systems, which were largely unprepared for remote visits and led to the disintermediation of providers. Direct-to-consumer products are now replacing the traditional doctor-patient relationship. It's clear that if healthcare providers do not adapt to offer digital health solutions, the industry will evolve without them.

Artificial intelligence (AI) is the new frontier and is already revolutionizing how care is provided, starting with office functions where it can alleviate some of the burdens of data entry. Ultimately, as AI matures, it will become the dominant force in the future.

MANAGEMENT OF CHRONIC DISEASE

Chronic disease management remains the "elephant in the room," posing the greatest challenge in controlling the cost of care. Eighty-six percent of the total cost of care is consumed in the management of patients with chronic disease. Traditional fee-for-service models fall short of addressing the holistic medical needs of the chronically ill. Hybrid payment models that combine per-member-per-month payments with fee-for-service align financial incentives with patient outcomes and represent the future of healthcare payment models.

CONSUMERISM

The days when the physician was in control of the doctor-patient relationship are over. Digital tools have allowed patients to take control of their own care. They can now obtain accurate information without going to the doctor. This has resulted in care being disintermediated away from the provider to third-party entities. Social media platforms are now the primary source of health information for most patients.

THE FUTURE OF GASTROENTEROLOGY IN THE ERA OF THE SECOND DERIVATIVE

What is the role of the gastroenterologist in this evolving landscape that I'm referring to as the era of the second derivative? What will healthcare and gastroenterology look like in the future? How will we adapt, and what must we do to sustain our practices? These questions have far-reaching implications because the fate of GI physicians is intrinsically linked to the broader industry. Shifts within this sector resonate throughout the entire healthcare landscape.

Praveen Suthrum has been a healthcare visionary for many years. In *The Shift*, his sequel to *Scope Forward*, he offers a forthright discussion about the ongoing disruption in healthcare through the lens of gastroenterology care. He courageously addresses challenging questions and provides critical analysis enriched with perspectives from the most brilliant minds in gastroenterology today. More than just a professional guide, this book reflects Praveen's personal journey and his visionary approach, which will motivate you to build a positive approach to change.

The Shift is a call to action—a prompt for all in healthcare to engage deeply with the forces of change and thrive in this era of the second derivative.

INTRODUCTION

Distraught that six pediatric visits couldn't alleviate her baby Annika's uncontrolled vomiting, Angelika Sharma had no choice but to quit her job and take care of the situation. After 18 months of Annika suffering severe food reactions, a Facebook ad pointed Angelika toward a Silicon Valley-based microbiome startup called Tiny Health. The company was offering a test for her child through a simple stool sample swab. A few weeks after sending in her daughter's sample, the results revealed that the baby's gut was loaded with a common bacteria called *P. vulgatus*. A Tiny Health nutritionist advised Angelika to give her daughter a probiotic and sauerkraut. They also prescribed an additional unusual solution—daily trips to a petting zoo to expose Annika to animal microbes. Following this regimen over a few months solved the problem. Annika was able to consume food normally, and new tests showed that her gut microbiome had changed.

When I read this story in a *Washington Post* article entitled "Doctors couldn't help. They turned to a shadow system of DIY medical tests" in June 2024, I felt it resonated wholly with the ideas I was engaging in this book. While the article debates the pros and cons of the home diagnostics market, my focus is on a larger trend transforming healthcare. Angelika and Annika's experience is not just a singular success story; it epitomizes a broader shift that's underway across healthcare—a shift enabled by digitization. The shift to digital.

At the turn of the 21st century, there were Internet companies, and then there were other businesses. Today, it's hard to imagine businesses outside of the Internet. Similarly, the coronavirus era laid the foundation of digital health, driven by the constraints of lockdowns and the urgent need for remote solutions. By the turn of this decade, all of healthcare will be enabled by digital technology, and it will be hard for healthcare organizations to stay relevant without embracing this shift. But this raises a critical question: Is the industry prepared for this digital revolution?

About a decade ago, Dr. Narayanachar S. Murali, a gastroenterologist based in Orangeburg, South Carolina, sent me an email after reading an article on the taboo of talking about doctors' suicides. He was moved. He listed several issues troubling physicians, including lack of control over professional activities, overload of mindless duties without reimbursement, debt burden, discordance between what society tells them a physician's life should be and what it really is, easy availability of drugs, lack of interests outside of profession and family, lack of spousal support, and severe loneliness. Like everyone in the story "The Emperor's New Clothes," we know that these issues exist. We just don't talk about them enough. Not openly enough. Despite the passage of time, the situation for doctors hasn't improved significantly. To add to doctors' woes, waves of innovation are transforming patient care, and many doctors are being left behind. If the principal actors of healthcare—our clinicians—aren't prepared, it poses a risk to the entire system.

But there is hope. There is light at the end of the tunnel. As the healthcare industry continues its relentless shift toward digitization, it has the potential to alleviate many of these long-standing systemic issues. Digital technology can help reduce the administrative burden on doctors. Innovations in AI can improve work-life balance by minimizing unnecessary tasks. Enhanced analytics can make treatments more personalized and effective. Digital platforms can offer better support networks for doctors. This shift to digital is not just about adopting new technologies but about creating a more sustainable healthcare environment. It needs clinicians who embrace change. It needs leaders grounded in technology. It needs you.

Gastroenterology, in particular, faces profound changes due to digital disruption. As digital innovation transforms patient care, the stakes have never been higher. If we do nothing, we'll be run over by digital disruption. But if we embrace the shift, we're looking at a different outcome. If you choose to act now, you can win on your terms. That's what this book is about.

In 2022, at an AI event conducted by MIT, the CIO of a billion-dollar digital health startup excitedly described a body scanner used to measure temperature and other vitals as patients entered their facility. The technology aimed to automate the collection of vitals, which AI analyzed to detect diseases proactively. While the idea seemed cool, I asked, "Why not use an ordinary thermometer instead of a life-size body scanner?" He responded that using a thermometer would involve manual input or a "human in the loop," thus reducing automation. Impressive as this sounded, I couldn't help but wonder, "So what? Aren't there more critical areas in which to innovate in healthcare?"

This example underscores a crucial point: While technological innovations can significantly enhance healthcare delivery, the role of real-world clinical experience remains indispensable. Clinicians bring critical thinking, context-specific insights, and ethical considerations essential to the shift to digital.

We've seen failures in the past when technologists overlooked the practical insights that clinicians provide. Consider the case of Proteus Digital Health, which developed ingestible sensors to track medication adherence. Once valued at $1.5 billion, Proteus faced skepticism and failed to gain market traction due to its complexity and high costs, leading to bankruptcy in 2020. Greater physician involvement might have helped steer the technology toward more practical applications, ensuring its success and acceptance in the market.

At the outset, many clinicians are wary of technology, believing it isn't their domain. But is that really true? Reflect on your work over the past year. Whether you've been choosing server systems, implementing telemedicine, or integrating AI into your practice, technology is already an inseparable part of healthcare. This engagement with technology is only going to multiply during this decade. In this evolving landscape, physicians need to reassess their relationship with medicine and the healthcare industry at large. This book will help you do that by providing insights into the future.

The Shift is a result of my long-held frustration with the healthcare system. On one side, I saw patients unable to access the right care at the right time. On the other, I saw brilliant physicians dissipating their time and energy in the business aspects of medicine. My frustration transformed into something more meaningful when I began to ask myself, "How can I be more useful?" That led to the founding of my company, NextServices, in 2004. Through our administrative work, I could see first-hand how the system had become this "giant sucking" machine usurping time, money, and resources by the billions. Today, my company is applying AI to automate the physician's back office.

This work led to ongoing interactions with gastroenterologists and my previous books, *Private Equity in Gastroenterology: Navigating the Next Wave* and *Scope Forward: The Future of Gastroenterology Is in Your Hands*. Since it was published in 2020, *Scope Forward* has deeply influenced the gastroenterology industry, changing the trajectory of many careers and organizations. On the heels of *Scope Forward*, I started a leadership program called the *GI Mastermind* to bring together future-oriented GI leaders. Participants often refer to it as an "eye-opening experience." In this book, you'll meet some of the real-life leaders who've taken huge steps in their entrepreneurial journeys.

My latest initiative is an in-person meetup called the *GI Mastermind Conclave*. In 2024, it brought together innovators and leaders in an immersive all-day workshop. What started as a way to help GI leaders survive the pandemic by learning from others became what's now called *The Scope Forward Show*. With 59 podcasts since 2020, it's become one of the most influential podcasts in the GI space. Combined with a curated weekly newsletter on the business/technology updates in GI and a website to research GI news, these initiatives continue to be an invaluable resource for the wider industry.

Through this long journey, I have learned a great deal from people innovating in the gastroenterology and healthcare space. This book is a synthesis of those learnings. It's where I see the world of GI and healthcare going. More importantly, it's a sincere appeal for the industry to embrace change in the face of world-changing digital events led by AI. The downside of inaction at this point is huge. It means disruption at a scale we haven't experienced yet—not just for gastroenterologists but the broader industry that the field supports.

When I wrote *Scope Forward*, my vision was for the GI industry to use the book to shift the course of gastroenterology. With this book, *The Shift*, I remain steadfast in that vision. This book goes further by identifying impending digital transformations and supporting them with real-world case studies that demonstrate that the future is already here. In essence, *The Shift* builds on the foundation laid by *Scope Forward*, pushing the boundaries further. It's an earnest call to action for the sake of the industry at large.

Chapter 1 (The Digital Shift: From Traditional to Transformative) examines the technological shift toward digital health. Chapter 2 (The Business Shift: From Linear to Exponential Models) explores how businesses can leverage exponential models to save costs and create new revenue channels. Chapter 3 (The Mindset Shift: From Burnout to Innovation) addresses the mental barriers we need to overcome to make these changes happen, and Chapter 4 (The Decisive Shift: From Watching to Shaping the Future) reimagines gastroenterology care in 2034.

I recommend reading the first section (chapters 1-4), highlighting the digital shift, end-to-end. Then, move on to reading about 30-plus trailblazing innovators. You may want to select and read those interviews that catch your eye first. You can always come back and read others later. The interviews are lightly edited and presented in a conversational style, ensuring you feel the authenticity and immediacy of each discussion. These leading voices cover a variety of topics: artificial intelligence, robotics, data, diagnostics, and devices. You'll also see innovation in other forms, from business to new care models. A few interviewees provide an open view of their careers that could help you reflect on yours. Finally, you'll get the industry's big-picture perspective from some of the leading voices in gastroenterology. The insights from the interviews are so practical and profound that you'll find yourself returning to refer to them again and again.

While this book celebrates innovation in gastroenterology, its themes are universal to all healthcare. It is aimed at leaders who wish to leverage this disruptive phase for transformative change. GI today is a hub of innovation. It's one of the hottest spaces to be in medicine. Whether artificial intelligence, robotics, or new care models, GI showcases it all. Getting a deeper perspective into this specialty can offer you great insights into the trajectory of healthcare itself.

To everyone engaging with this book, I hope you will take up the challenge to lead and influence. Whether you're a solo practitioner, part of a group practice, engaged in a private equity organization, working at an academic center, an administrator, a digital health entrepreneur, a med-tech company, an investor, a consultant, an executive in big pharma, or even a patient, use this book to embark on the path of influence. What's crucial now is the determination to drive change by positively influencing those around you. All of us have a choice: Either we shape the system for the better or allow the system to shape us.

Instead of deliberating on the change, let's embrace the shift and make it happen. We are lucky to be here right now to watch history unfold. It's important to understand the direction in which the industry is headed and play our part in it. We need to ensure that we aren't left behind in a world that's changing at unprecedented speed. This is our big opportunity to reimagine gastroenterology care and positively impact patients' lives at a scale unimaginable at any other point in history.

If this message resonates with you, share this book with colleagues and peers who could also benefit from adapting to the shift. Now is the time for all of us to grasp the opportunity this transformative era in gastroenterology and healthcare brings. Most importantly, let the stories I've shared here inspire you to engage with technology, *stop stopping yourself,* and take action.

A QUICK NOTE ON PROFESSIONAL ADVICE

This book is not intended to provide financial or legal advice or steer you in a particular direction. It won't answer all your questions. Rather, it'll help you ask the right questions to make decisions that work for you. All opinions expressed belong solely to me as the author and do not belong to the organizations I may be associated with. This book isn't a substitute for professional expertise from attorneys, accountants, or investment bankers.

SECTION: AN IRREVERSIBLE SHIFT

This section examines the transformative impact of digital innovation on healthcare, from foundational changes and exponential business models to innovative mindsets and shaping the future.

Chapter 1: The Digital Shift: From Traditional to Transformative

- Explores the foundational changes brought by digital health technologies, accelerated by the COVID-19 pandemic, and sets the stage for understanding the digital transformation in healthcare.

Chapter 2: The Business Shift: From Linear to Exponential Models

- Discusses how healthcare businesses can leverage exponential models to innovate and thrive in a rapidly changing digital landscape.

Chapter 3: The Mindset Shift: From Burnout to Innovation

- Highlights stories of innovation and resilience in gastroenterology, addressing the shift in mindset needed to embrace digital transformation and overcome professional burnout.

Chapter 4: The Decisive Shift: From Watching to Shaping the Future

- Reimagines the future of gastroenterology and healthcare, emphasizing proactive leadership and the impending AI revolution.

CHAPTER 1

THE DIGITAL SHIFT: FROM TRADITIONAL TO TRANSFORMATIVE

It was the end of the summer of 2007. As I rode my motorcycle across the high-altitude desert plateaus of the Himalayas, the world seemed distant and inconsequential. Despite holding an MBA from a top US school, I was struggling to cover my monthly student loans.

In the US, the Great Recession, triggered by events like the housing bubble collapse, was around the corner. The world was at a critical juncture, facing an economic downturn and widespread unemployment. But I was oblivious to these shifts.

At that point in my life, I was ensnared in a trifecta of personal challenges connected to money, business, and a relationship. The uncertainty of each payroll cycle in my startup kept me on edge. To compound matters, a personal relationship was a tangled mess. Regardless of what I did, nothing seemed to kick into gear. I needed a pause.

My radical response to this life situation was to get on a motorcycle and drive up to Ladakh in the Himalayas. It wasn't just another road trip up to the mountains. I embarked on a solo expedition to Khardung La, then the world's highest motorable road. It was a daunting 1,500+ mile trip from Mumbai, where I was situated. It was crazy. But crazy was exactly what I needed.

At times, I questioned why I was navigating through this barren landscape. Each day, I drove with my fingers crossed, hoping to reach a campsite before dusk. Hours would pass without spotting a living thing. Sometimes, I'd spot the remains of an animal. Occasionally, I'd see a little shrub or two that had managed to grow despite the bare, arid landscape. Ladakh is a dry, cold, unforgiving desert in the rain shadow of the Himalayas.

One day, I was driving on a straight road through sandy terrain. In the distance stood a surreal Himalayan range. Cold wind blasted my face as I felt encapsulated by the mythical beauty of the landscape. Suddenly, I lost control of the motorcycle and fell.

My motorcycle had skidded on the sand, and I had taken a bad fall. The side stand broke. My belongings were scattered around me. I was physically alright, but the thin Himalayan air made each breath a struggle. I hadn't seen a single soul for several hours, and mobile phones didn't work at that altitude back then. Getting help was going to be impossible. This wasn't how I wanted my road trip to end.

As I gathered myself after the fall, the image of a dead horse's hoof vividly flashed in my mind. I had noticed it a few hours earlier, striking against the dry terrain. The low oxygen and air pressure levels made my head spin. I sat down, took a few breaths, and waited. Once I felt steadier, I picked up my gear. Using my back, I slowly raised the motorcycle—a Royal Enfield Bullet, weighing three times my weight. I prayed it would start. It did. As I got back on my motorcycle, I saw that one of my gloves had fallen in the distance. For a moment, I shuddered, remembering how cold my hands would get during the icy cold nights in the mountains. I considered the effort of laying the motorcycle down and picking up the glove. I decided to give it up and drove on.

In those moments, I began to understand why I had undertaken this arduous journey. Though I couldn't articulate it precisely, I could feel a deep sense of clarity on what I *didn't* want for my life. I was trying to meet my family's expectations of a "good life." My business had become a facade adhering to society's norms of success. Personally, I found my true self buried beneath layers of people-pleasing behaviors. I realized I had lost my way. And yes, it took a wild solo road trip on a motorcycle to stumble upon that insight. That journey marked a profound shift in my world.

Just as I was experiencing a personal transformation, the world around me was also undergoing a significant shift. A subtle yet powerful digital transformation was beginning to take root, reshaping industries and societies in ways we had never imagined.

IN 2007, THE WORLD SHIFTED TO DIGITAL

The period from 2006 to 2009 saw the onset of the Great Recession. The US housing bubble burst largely triggered the financial crisis, causing widespread financial instability. The International Monetary Fund considered this crisis the most severe since the Great Depression of the 1930s. The unemployment rate had doubled to 10 percent. Home prices fell 30 percent. And the S&P 500 stock index was down 57 percent.

It was hardly a period you'd associate with innovation. Yet, something was brewing. Society was steadily shifting to a digital-first approach. Computers shrunk. Phones got smarter. We were learning to find many new reasons to be online.

In 2006, Google acquired YouTube for $1.65 billion. That same year, Spotify, a music streaming service, was founded in Stockholm. Twitter (now called X) started its 140-character blogging service that year.

In 2007, Apple launched the iPhone, a groundbreaking device that combined a mobile phone, an iPod, and an internet communicator, featuring a revolutionary touchscreen interface and a full web browser. That was followed by Google launching the Open Handset Alliance to develop the Android operating system. Amazon introduced the Kindle e-reader. Facebook opened itself beyond universities. Two former roommates, Brian and Joe, turned their San Francisco apartment into a bed and breakfast and started Airbnb. Netflix began delivering movies to "the PC." IBM launched the Watson project that famously won Jeopardy! in 2011.

That same year, genome sequencing, the process of determining the complete DNA sequence of an organism, began to drop steadily from $350,000 to below $1,000 by 2020. The National Institute of Health launched the Human

Microbiome Project, marking a significant milestone in recognizing the role of the microbiome in gut health.

The following year, in 2008, cryptocurrency got a start through a paper written under the pseudonym Satoshi Nakamoto. Microsoft, which had until then missed out on the internet race, launched Azure, a cloud computing service. Meanwhile, apps started proliferating via the Apple App Store. A year after that, in 2009, former employees of Yahoo started WhatsApp, which was eventually acquired by Facebook for $19 billion five years later. And, of course, Travis Kalanick cofounded a timeshare limo service called Uber.

THE ADVENT OF DIGITAL BIOLOGY

In March 2008, amidst these global changes, a short but significant article in the research section of *The New York Times* may have gone unnoticed by many. It highlighted the inclusion of two tests for colorectal cancer prevention and detection: a virtual colonoscopy and a stool DNA test to find abnormal DNA associated with cancer. This marked a pivotal moment in the medical field, signaling a shift toward non-invasive diagnostic methods.

With reference to the stool DNA test, Dr. David Ahlquist and colleagues from the Mayo Clinic published the results of their research, proposing the test as a new approach to colorectal cancer detection. It would work by analyzing DNA from a stool sample to detect abnormal genetic markers associated with cancer cells. This method allowed for early detection of cancer without the need for a colonoscopy. Curiously, their paper says in its background that "few data are available from the screening setting."

They didn't have to wait too long for the data to multiply. Exact Sciences collaborated with Mayo Clinic to develop and commercialize Cologuard®, a stool DNA test. The FDA approved Cologuard® in 2014, and the Centers for Medicare and Medicaid (CMS) began to cover it the same year. Soon, data from Cologuard® tests began to grow exponentially.

To grasp the context of how explosive this growth has been, we need to fast-forward the story to the present. In a 2023 earnings call to investors, Jake Orville, General Manager of Screening for Exact Sciences, described these statistics. In 2015, one year into Cologuard®'s launch, they performed 100,000 Cologuard® tests. In the 11 days before that day's call, they had performed 100,000 Cologuard® tests.

It's hard to miss the exponential *shifts* leading to this narrative. Doesn't it feel like we've *always* used the iPhone, hailed an Uber, tweeted on X, and streamed movies on Netflix? Yet we know it wasn't *that* long ago when none of it was possible. These shifts and the swift growth in digital solutions are never linear; they are *exponential*. The exponential shift to digital is so gradual and subtle that it doesn't feel sudden.

But it wasn't just about digital technologies. Stool DNA testing exemplified digital biology, which applies computational techniques for more accurate, non-invasive diagnostics. My previous book, *Scope Forward*, alluded to the upcoming surge in stool DNA testing and AI. When the book was published in 2020, Cologuard® had already screened 1.7 million people for colorectal cancer using its innovative stool DNA test. By 2023, this number had skyrocketed to 12 million. This rapid increase in screening highlights the growing acceptance and reliance on non-invasive diagnostic methods in healthcare.

When you view stool DNA testing as "digital biology," it's easier to see the turn that colon cancer detection could take. With more data, the tests would be more accurate and efficient. *Digitization* would place the field on an exponential curve (more on this shortly). In 2020, the gastroenterology industry was fighting Cologuard® and arguing that a traditional colonoscopy was still the "gold standard." While colonoscopy remains the gold standard as I write this, the industry no longer doubts the role of non-invasive tests. In fact, there is now a proliferation of non-invasive approaches to detect cancer.

The industry showed similar resistance when computer vision AI tools were introduced to assist with polyp detection. AI has the potential to catch polyps that may be missed by the human eye, resulting in more accurate diagnoses of colorectal cancer. Initially, many skeptical gastroenterologists used AI in endoscopy to check their own diagnostic assessments and confirm traditional methods as the "gold standard." However, at least one practice now refers to the utilization of AI in polyp detection as the "Platinum Standard."

Strong resistance to technological shifts reveals underlying fears within an industry. After all, didn't 6,000 math teachers stage a protest against the use of calculators in 1986? They carried signs that said, "Beware: Premature Calculator Usage May Be Harmful to Your Child's Education." Non-invasive tests, AI, and other digital technologies put the present-day livelihoods of gastroenterologists at risk. The inherent resistance to these shifts is understandable. Looking at it differently, it reaffirms that the shift is real.

Gastroenterology, and possibly all of healthcare, is shifting to digital. And *when a field goes digital, there's no turning back the clock*. It'll continue to expand digitally and latch onto an exponential curve. Once it latches onto an exponential curve, it'll appear deceptively slow initially but multiply over time to disruptive proportions.

We will examine why that's the case shortly. But first, we need to analyze the striking parallel between the pre-Great Recession era (2006-09) and the COVID-19 pandemic era (2020-23).

COVID-19 LAID THE FOUNDATION FOR DIGITAL HEALTH

Just as the years leading up to 2009 were a crucible of digital innovation, birthing technologies that would redefine our social and personal landscapes, the COVID-19 period served as a similar catalyst for healthcare. The necessity for remote interactions and the urgency of health monitoring during the pandemic propelled telemedicine, wearable tech, non-invasive tests, and AI diagnostics into the mainstream, mirroring the earlier tech boom's impact on communication and media.

To draw an even sharper parallel, consider the iPhone's launch, a harbinger of the smartphone revolution. Similarly, the spread of non-invasive diagnostics, such as stool DNA tests and AI tools, during the pandemic signaled a shift toward more accessible and patient-friendly healthcare. Both instances exemplify how something suddenly propels forward after brewing under the surface for a long time.

Moreover, the proliferation of digital platforms and tools during the COVID-19 period mirrored the explosive growth of apps in the wake of the iPhone's debut. Healthcare apps and platforms have become the new 'apps' post-COVID, signaling an irreversible shift in how healthcare services are delivered and consumed.

The key takeaway is that these periods of intense innovation are not mere coincidences but are driven by pent-up societal needs that demand rapid advancement and adoption of technology. The years 2006-2009 set the stage for a digitally interconnected world, while the years 2020-2023 did the same for a health-focused digital transformation.

To truly grasp the transformative currents sweeping through healthcare, it's important to zero in on a specific domain that epitomizes these changes. My focus lies with gastroenterology (GI). As you will see, it's a field that has become a hotbed for digital innovation.

If gastroenterology is navigating toward a digital future, it signals a wider shift within the healthcare industry, suggesting that the same digital forces revolutionizing GI are set to redefine healthcare at large.

GASTROENTEROLOGY IN THE MIDST OF A MASSIVE WAVE OF INNOVATION

During the COVID-19 pandemic from 2020 to 2023, the field of gastroenterology witnessed a significant wave of innovation that altered how patient care is delivered. It was a transformative era in gastroenterology, driving advancements that promised to reshape patient care well beyond the pandemic. This period highlighted the efforts of companies like Geneoscopy, Iterative Health, Oshi Health, and Virgo

Surgical Video Solutions, all of which were established before COVID-19.

Before the pandemic, Geneoscopy had already secured $6.9 million to enhance colorectal cancer screening technologies. By 2021, the company had raised an impressive $105 million to further develop its non-invasive RNA screening test. Iterative Health, on a parallel track, raised $13.5 million in January 2020 to develop artificial intelligence tools tailored for gastroenterology. This momentum continued with a $30 million funding round in August 2021, followed by an additional $150 million by January 2022.

Oshi Health, aiming to revolutionize virtual gastrointestinal care, raised $23 million in October 2021 and an additional $30 million in April 2023. Meanwhile, Virgo Surgical Video Solutions, established in 2017, adapted its focus during the pandemic toward screening patients for clinical trials. Significantly, it became the first company to receive investment from Olympus Innovation Ventures, the venture funding arm of Olympus Corporation, known for its endoscopy systems.

When I asked Dr. Erica Barnell, cofounder of Geneoscopy, why investors were attracted to her company, she said, "While I believe our concept is excellent and has the potential to save lives, it's our execution that really draws investors. They've noticed how well we adapt to changes in the environment and how effectively we can pivot in response to external challenges, leveraging our resources for success. Investors seem to place their trust in teams that demonstrate resilience and adaptability, especially when unforeseen circumstances arise, such as the COVID-19 pandemic."

"For example, we had to conduct a clinical validation study during the pandemic, which presented a set of challenges nobody could have anticipated," she continued. "There was no playbook for conducting such a study under these conditions. We had to innovate and find novel ways to recruit patients for the study. Impressively, not only did we manage to do this more cost-effectively compared to others conducting similar trials, but we also completed it in 12 months, which is exceptionally fast. This ability to be creative, resourceful, and resilient has been crucial in gaining investor support and belief in our mission."

As of this writing, Geneoscopy is raising another $100 million in Series C funding.

This shift isn't just about funding digital GI startups. The pace of innovation across gastroenterology accelerated during COVID-19. The constraints propelled gastroenterology innovators to think differently and showed newer possibilities.

FoodMarble, for instance, introduced a second-generation breath analysis device to diagnose gastrointestinal disorders. Centaur Labs leveraged crowdsourced polyp classification for clients developing AI tools. Alimetry's wearable technology was cleared by the FDA, allowing the non-invasive electrical monitoring of gut activity. Dieta Health raised a significant funding round to develop AI-powered stool image recognition for smartphones, contributing to more accessible digestive health management.

Medtronic made significant strides with the launch of GI Genius™ in the US, the first AI-based system for polyp detection during colonoscopies. Endosound won an award at the

AGA Shark Tank and worked to democratize endoscopic ultrasound with innovative add-on devices. Each of these developments underscores a broader trend.

Broadening our view, digital health funding experienced a dramatic increase, rising from $7.9 billion before the COVID-19 pandemic to $29.1 billion in 2021. This surge in investment reflects the growing importance and adoption of digital health technologies. These investments are reshaping the healthcare landscape by making healthcare more accessible, efficient, and personalized. It's naïve to think that this influx of capital won't significantly impact healthcare.

With the foundation for digital gastroenterology laid, we're looking at a sweeping transformation of healthcare itself. How will traditional medical practices or organizations deeply entrenched in a specialty compete with the relentless tide of digitization? As healthcare leaders at the crossroads of these transformative currents, how can you harness and tackle these challenges?

MARKETS EVENTUALLY REVEAL THE FUTURE

Before we address these questions, let's briefly go back to digital innovations during the pre-Great Recession era and see how they played out in the stock market—particularly NASDAQ, an index skewed toward tech stocks.

When you observe the index from 1990 to 2002, you can see the peaks of the dot-com boom and then its bust. The index rose again until 2007, busting again during the Great Recession.

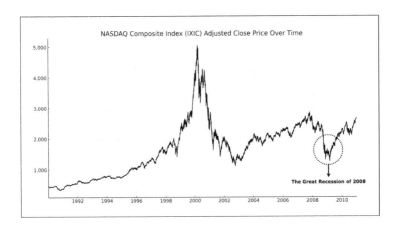

At that time, when NASDAQ hit a low, you would've thought that the tech wave was likely a bubble. And we'd never see another peak like the dot-com era. However, something else happened. Post 2008, NASDAQ experienced historic growth, surpassing all previous benchmarks.

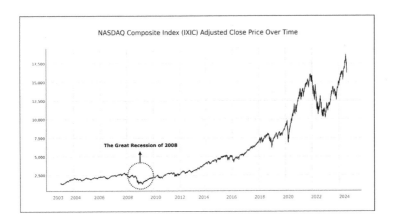

Take a look. The Great Recession looks like a minor blip in NASDAQ's story.

Returning to examining the COVID-19 period, the stock markets hit a peak in 2021 and nosedived in 2022-23 when digital funding dried up. But then, led by AI innovation, by the end of 2023, the index began to rise again to newer highs.

To assume that digital health would not be sustained beyond COVID-19 is akin to believing in 2002 that the dot-com era was over or in 2008 that tech is all a bubble. Markets constantly adjust themselves but follow long-term economic and technological trajectories.

Today, we are amid a massive wave of disruption in healthcare. It's shifting the field from the *scarce* domain of doctors and patients to an *abundant* realm fueled by digital changes.

A SHIFT FROM SCARCITY TO ABUNDANCE

The big difference between the Tesla Cybertruck and the GMC truck isn't just that one is autonomous and electric, and the other is not. While the GMC truck has an excellent computer inside it, the Tesla is, in reality, a *sophisticated computer on wheels*.

To grasp what that means, we need to get back to basics. We need to understand what "digital" means exactly. And why shifting to digital has the potential to plug a field onto an exponential curve.

Semiconductors are the bedrock of computing and digitization, and this digitization paves the way for exponential growth. These tiny components are essential in the manufacturing of microchips and processors that power computers, smartphones, and other digital devices. As semiconductor technology advances, it enables more powerful and efficient computing capabilities, driving exponential growth in various digital applications. Once a field transitions to digital, be it gastroenterology or medicine at large, it enters a phase of exponential growth driven by continuous advancements in computational power and data analytics. It becomes subject to the same rapid technological evolution we've seen with smartphones and computers.

Such a shift is largely due to the field starting to mirror Moore's Law. The law posits that the number of transistors on a microchip doubles approximately every two years, leading to a corresponding increase in computing power. As computing evolves, it enables faster, more accurate diagnostics and personalized treatment plans, transforming the healthcare landscape at an accelerated pace.

In 1965, engineer Gordon Moore published a seminal paper in *Electronics Magazine* titled "Cramming more components onto integrated circuits." That cramming forecasted by Moore's Law, far from being a short-term trend, has held true for over half a century. Today, we see companies like Graphcore integrating 59.4 billion transistors and 1,500 processing units onto a single silicon wafer.

The implications of Moore's Law extend beyond technology-centric industries, reshaping the entire digital world. For perspective, electricity took 46 years to reach 50 million users, mobile phones 12 years, the internet seven years, and Facebook four years. In stark contrast, Open AI's ChatGPT reached one million users in just five days and 100 million users within two months of its launch in November 2022. As of August 2024, ChatGPT has over 180 million users.

Smartphones are another example of exponential growth. From their inception in 1996, it took 16 years to reach one billion users. Then, in a mere four years, that number doubled. By 2022, an astonishing 83.7 percent of the global population, or 6.64 billion people, were using smartphones.

It's now healthcare's turn to harness the power of Moore's Law.

A RECAP ON LINEAR VS EXPONENTIAL TECHNOLOGIES

In the first chapter of my book, *Scope Forward*, I explored the distinction between linear and exponential technologies in the medical field. Let's simplify that discussion to understand the fundamental differences more clearly.

Humans are historically inclined toward linear thinking, perceiving change as a series of small, incremental steps. This viewpoint suggests that developments happen gradually, in a predictable and steady fashion, mirroring a one-to-one scaling relationship. For example, a gastroenterologist might think, *My colonoscopy calendar is full; what could possibly disrupt that?* Essentially, we expect tomorrow to be a slight modification of today.

Contrastingly, the evolution of various industries showcases the dominance of exponential growth, particularly in areas fueled by digital innovation like digital biology, robotics, 3D printing, autonomous vehicles, blockchain, and more. These fields don't just grow; they explode in capability and impact, doubling performance or capacity at regular intervals. This type of growth defies the linear model, adopting a pattern where advances multiply rapidly, illustrating a one-to-many or many-to-many expansion rate.

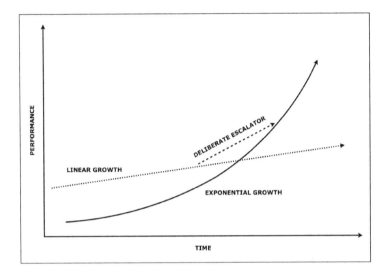

The real challenge for us is that our brains are wired to understand linear growth, making it difficult to comprehend the nature of exponential progress intuitively. We tend to extrapolate from the past and expect digital advancements to follow a similar, steady trajectory. However, the reality of digital progress is more akin to a series of doublings—one becomes two, then four, eight, sixteen, and so on, escalating far beyond linear predictions.

To truly appreciate the digital revolution's pace and impact, we must consciously adjust our perspective, stepping onto a "deliberate escalator" that cuts our linear mindset and shifts us to an exponential one. That will allow us to recognize and engage with the nature of today's technological advancements.

This point merits emphasis: *When a field transitions to digital, it starts to follow Moore's Law. If gastroenterology shifts to digital, it'll follow Moore's Law. If healthcare itself shifts to digital, it'll follow Moore's Law. From that point on, there's no turning back.*

As we have examined so far, the shift to digital is inevitable in healthcare, just as it is in all major industries. However, our future can be scary or exciting based on how we look at it. As a healthcare leader, there's a key element that you need to pay attention to. And that has nothing to do with market or technological conditions. It's simply to do with you alone. Let's examine what I mean.

WHEN I STOPPED STOPPING MYSELF

When I fell off my motorcycle in Ladakh, I was spooked mentally. I had a choice to make. To return and go back home to comfort. Or to go forth and lead myself to a different future. I needed to overcome fear, listen to my inner voice, and get on with building my life on my terms.

It wasn't that I knew how the journey would play out. I didn't. My motivations were *because* I didn't know, and I was curious to see what would happen if I stepped forward. That moment of risk was my Shackleton moment (more on this soon). After that day, I simply *stopped stopping myself* and went about taking action on things that mattered. I began to accept that in my heart, I was a thoughtful artist and an adventure-seeking explorer and not a Jack Welch-style businessperson. What I build must reflect who I am. Before I related to others, I had to learn how to relate to my true self. I had to learn to master my emotions, especially those that triggered fear and put me off balance.

That fall off my motorbike and my reaction to it triggered a change in how I viewed the world. In fact, that trip changed my whole world.

While a motorcycle journey was my thing, it may not appeal to many others—I get that. The point isn't about undertaking risky journeys; it's about finding *your* Shackleton moment. As a healthcare leader, it's important to recognize that this time of rapid innovation and digital disruption is a historic juncture. Beyond technological or business reasons, you have a personal decision to make right now. *You* are at the crossroads of an opportunity that can be life-changing and industry-shifting. Let's explore this thought a little further.

THIS IS A SHACKLETON MOMENT

In 1914, Irish explorer Ernest Shackleton embarked on a daring Antarctic expedition to cross from the Weddell Sea to the Ross Sea, an extremely daring and ambitious exploration challenge. The expedition's ship, *Endurance*, became trapped in pack ice in the Weddell Sea and was slowly crushed, forcing the crew to abandon it. Marooned on ice floes, they endured harsh conditions for months. Shackleton and five others eventually undertook a perilous 800-mile journey to South Georgia Island in a lifeboat. After an arduous trek across its mountainous terrain, Shackleton finally secured rescue for his crew. They had been through more than two years of extreme endurance, but all crew members survived.

As a leader, Shackleton played a crucial role in maintaining morale and hope. He kept the crew engaged in tasks, structured routines, and leisure activities to fend off despair and maintain physical and mental health. He was a hands-on and inclusive leader who worked alongside his team.

Shackleton's story is not just one of survival; it's a narrative on overcoming the impossible, a principle that resonates deeply with those of us committed to pushing the boundaries of what is achievable.

As healthcare leaders at the crossroads of digital disruption, we are at our own Shackleton moment. For some, this shift evokes frustrations akin to what the industry experienced with the electronic health record (EHR) system—seen as an unnecessary burden and even further complicating the simple act of a doctor seeing a patient.

That mindset, driven by a sense of contraction and fear, is that of *scarcity*. It makes you want to grab what you have and checkout. You'd rather sell out and leave the game.

Yet, what if we viewed digital disruption not as a hurdle but as an inevitable wave of progress? A Shackleton moment that'll redefine healthcare. What if you view this disruption as an adventure and challenge to deal with head-on? That shift in mindset will suddenly help you look at the shift to digital as a lifetime opportunity that will change the entire game.

It was psychologist Carol Dweck who showed through her research that individuals with a growth mindset tend to believe in the abundance of opportunities. In contrast, those with a fixed mindset perceive limitations and scarcity.

It's time for you to choose what state of mind you wish to be in on this journey. It's not because success is guaranteed or that the path will be easy. But it's because the very challenge makes you come alive.

Should you wish to go on this adventure, adapt to the mindset of abundance. Use the coming decade as your playground to reshape the destiny of not just your career but the entire industry.

Right now, you stand at a pivotal juncture. You have a choice to make: Will you lead and harness the digital wave to redefine healthcare, or will you watch from the sidelines as the future unfolds without you? We'll revisit this critical decision in Chapter 4, exploring the implications of these changes in gastroenterology and healthcare. For now, we need to recognize that this wave isn't just technological; it's also about rethinking your business model. In the next chapter, we'll explore the shift from linear to exponential models and how they help businesses stay competitive.

CHAPTER 2

THE BUSINESS SHIFT: FROM LINEAR TO EXPONENTIAL MODELS

"The best support I ever received was a pamphlet and instructions to go figure it out on my own," said Sam Jactel, now CEO of Ayble Health, about managing his severe ulcerative colitis. Despite receiving top-tier medical care, he found himself unable to access all the evidence-based therapies discussed at his appointments. The lack of easy access to recommended therapies, particularly dietary and psychological, left him in a frustrating position.

Following his diagnosis, Sam consulted gastroenterologists at world-renowned medical centers for over ten years, experimenting with various drugs in search of an effective treatment. Despite accessing top-tier care, he still experienced significant flares because he was unable to access evidence-based lifestyle and behavioral care – a frustrating position to be in, given those therapies are an integral part of clinical guidelines from gastroenterology societies. Bedridden with a severe flare in 2020, he realized he needed to take control of his own care. He couldn't rely on the existing healthcare system.

Sam realized there was no easy way to access evidence-based recommendations that his physicians suggested, such as personalized dietary or psychological therapy. He soon began working with researchers at Massachusetts General Hospital and Northwestern Medicine to create a personalized nutrition and behavioral health program that eventually turned into over a dozen peer-reviewed publications. Sam didn't stop there. As a Kellogg-educated MBA, he figured that if he could systematize protocols using a care team and AI, he could help other patients with digestive disorders like his. That endeavor led to the founding of Ayble Health, a patient-centered company that helps people take control of their gut health.

Sam's frustration with the inadequacies of the healthcare system transformed him from a patient to the CEO of a venture-funded digital health startup. He says, "You buy a pair of Nikes, and Nike sees that grey shoes with a red stripe are popular; they produce more of that kind. In healthcare, there's no direct way for patients to vote with their dollars. Patient needs are either filtered or not heard at all."

In his words, "Being patient-centered is different from being patient-driven." His story reinforces the message that a patient-driven organization pulsates with a deep sense of purpose.

Purpose-driven digital health companies aren't just the domain of frustrated patients. They are also the domain of frustrated physicians.

Growing up in Singapore, Dr. Jonathan Ng spent 14 years making frequent trips to Cambodia, where he played a crucial role in establishing the country's healthcare system. His efforts included setting up a range of facilities, from cardiothoracic units to pediatric hospitals in rural areas.

Jonathan's team brought in medical educators from Singapore or the US to train doctors in Cambodia. He recalls, "A recurring challenge we faced was in diagnosing conditions. I remember an instance in the operating theater where my mentor pointed out a massive tumor, describing its details and treatment options. I could see it, but the local surgeon we were training couldn't even locate it."

That episode led to a profound realization. If doctors can't even identify the medical problem, how will they treat it? A different solution was needed that could help doctors with limited training diagnose conditions. He later wrote to me, "They'd basically whip out a textbook and try to learn in real time. Doesn't quite work that way."

After more than a decade, he could see that he was spinning in circles trying to change healthcare. The progress was too slow. He needed a new approach. Taking a break, he went to the Massachusetts Institute of Technology (MIT) to study business and Harvard Kennedy School to study healthcare policy. The educational stint catalyzed his startup, and he didn't feel the need to complete it.

Keeping an open mind helped him spot something unexpected. Computer vision for autonomous vehicles was being used to spot cats, dogs, and humans. He thought, *Why can't* we use it to provide equitable healthcare outcomes? What if you used AI to spot something like tumors?

That insight led to the formation of Iterative Health (formerly Iterative Scopes), which is redefining gastroenterology care today by building AI tools to detect polyps in the gut.

Once again, it's impossible to overlook the deep sense of purpose in Jonathan's reason for building his startup.

MASSIVE TRANSFORMATIVE PURPOSE

Transitioning from these personal stories, we see that the concept of Massive Transformative Purpose (MTP) is at the heart of an exponential model. MTP is not just a vision statement crafted after a company's formation. The MTP is its raison d'etre, its very reason for its existence. It's about envisioning how your organization can make a monumental difference in the world. The key here is "massive"—how can your organization make a massive transformation happen?

Let's look at some examples of MTPs from renowned companies:

- **Tesla**: Accelerating the world's transition to sustainable energy.
- **Virgin**: Changing business for good.
- **Google**: Organizing the world's information to make it universally accessible and useful.
- **Unilever**: Making sustainable living commonplace.
- **TEDx**: Spreading ideas worth sharing.
- **Netflix**: Entertaining the world.

- **Starbucks**: Inspiring and nurturing the human spirit.
- **Spotify**: Unlocking the potential of human creativity.
- **Nike**: Bringing inspiration and innovation to every athlete in the world.
- **Patagonia:** We're in business to save our home planet.
- **Airbnb**: Creating a world where anyone can belong anywhere.
- **Ikea**: Creating a better everyday life for many people.

These MTPs amplify the mission of each organization. Patagonia, for example, doesn't simply aim to make sustainable clothes—their mission is to save the planet. Starbucks isn't just about brewing world-class coffee; they seek to nurture the human spirit.

Of course, no single organization can achieve its MTP in isolation. That's the essence of these goals—they are comprehensive and all-encompassing. Guided by their MTPs, businesses can structure themselves, forge partnerships, attract talent, and draw in customers, all contributing to nudging the world in the direction of their overarching purpose.

Consider your work this past week. What's your purpose for doing what you do? At your very core, *why do you do what you do?* Let me demonstrate the power of purpose with my own story.

"You came out of nowhere."

A common quip I hear in gastroenterology circles is, "You came out of nowhere." Most people are surprised to learn that I'm not a clinician, let alone a gastroenterologist. Then they stare in disbelief when I quietly share that I don't live in the US but influence the GI space in the country from afar.

A few years ago, GI doctors would have recognized me as an uncertain entrepreneur selling software or billing services at conferences. I liked what I did, but there was this nagging feeling that I wasn't tapping into my full potential. I had much to share with the world but didn't know how.

What changed wasn't market conditions but me incessantly asking myself *what I really wanted* and *why*. What's my purpose, my MTP? Eventually, I realized that the purpose that deeply resonated with me at the time was "Enabling healthcare as a force for good." I was frustrated with the problems that plagued physicians and patients. Physicians sought patients to sustain their practice, and yet patients struggled to access care when they needed it. Instead of using money as a tool to improve care, economic incentives in healthcare sometimes encouraged unnecessary treatments. Additionally, doctors were largely unaware of the impending disruption from advanced technology. To make matters worse, doctors were burnt out and tired of the healthcare system. Something had to change.

In the beginning, I felt awkward sharing my MTP with others. But the more I said it, the more real it became to me. The more I *felt* my MTP in my bones, the easier it became to take massive action. I wrote books like *Private Equity in Gastroenterology* and *Scope Forward* that have struck a deep chord with the GI community. During the COVID-19 period, when everyone was worried about losing their livelihood, my MTP propelled me to start a podcast to help GI doctors not just survive the pandemic but *scope forward* to the future.

Today, *The Scope Forward Show* is one of the most popular business/tech podcasts in gastroenterology.

On the heels of that, I started a leadership program called GI Mastermind. The program offers deep insights into the wave of digitization in gastroenterology. Those insights helped some of the most well-regarded GI leaders and executives optimize business models and develop new leadership perspectives. For some, it led to personal transformation and career changes.

Learnings from the podcast and GI Mastermind led to the book you are currently reading. And I'm not stopping here. Through digitization, I plan to make GI Mastermind available to the entire industry so that the field of GI isn't caught off guard by the disruption that's underway.

Where did my journey begin? With a simple MTP that mattered to me. Do you see how powerful *purpose* is? It has the capacity not just to change you but also to create a momentum that can be a force in itself.

Determining your aspirational personal or organizational MTP is the crucial first step to developing an organization with exponential growth. To get to that, let us first understand the concept of exponential models.

UNDERSTANDING EXPONENTIAL MODELS

The ideas for exponential business models are well explained in the book *Exponential Organizations* by Michael Malone, Salim Ismail, and Yuri van Geest. Here is a simplified version of those concepts.

At the outset, embracing all aspects of exponentials isn't necessary. Choose what's important for your business today and then expand. However, placing your purpose at the core of all your decision-making is essential.

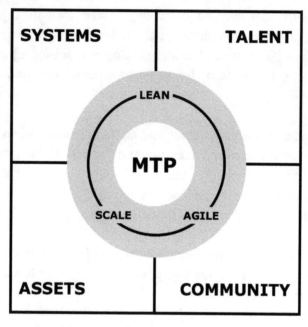

Adapted from the concepts in *Exponential Organizations* by Michael Malone, Salim Ismail, and Yuri van Geest.

Exponential models must embody three key attributes at their core: they need to be *lean, agile,* and *scalable* (surrounding MTP). Being **lean** involves creating efficient operations that optimize resources and minimize costs. **Agility** means being nimble and capable of quick iteration and fast failure to find the optimal path. **Scalability** requires thinking beyond current limitations. For instance, not just catering to 100 patients a day but envisioning a model that can efficiently serve at least 1,000 (10x).

The four key pillars that support exponential models are Systems, Talent, Community, and Assets.

Systems: For exponential growth, robust systems that help you automate workflow and scale are crucial. For example, several years ago, my team at NextServices performed medical billing activities manually. A specific process, such as checking the status of submitted claims or procuring benefits associated with a patient's insurance, would take an inordinate amount of time. Most of the workflow has now been automated using tools such as Microsoft PowerAutomate, a robotic process automation tool. Today, a team member "watches over" algorithms that perform many of the tasks and only intervenes when necessary. Areas such as pathology billing are nearly fully automated. For certain insurances, the team has successfully automated the procurement of prior authorizations, approvals required before specific procedures can be performed. The benefits are saving resources and time, predicting quality, and providing a foundation for scale. Now, the company is centralizing AI-driven processes for revenue cycle management (RCM).

Talent: Most of the talent you need lies beyond the physical boundaries of your organization. A linear mindset limits your ability to access talent from wherever it's available. For instance, a microbiome ancillary business might require expertise ranging from data science to legal consultation, often found through partnerships, consultancies, or even AI, rather than traditional hiring.

When a startup I was involved in needed a logo, we knew hiring a high-quality designer would cost over $2,000. Thinking more exponentially, we ran a competition with 99designs, tapping into a global pool of designers. We got 200+ logo submissions for $800 from premium designers in a few days. Finally, we chose a designer who refined the idea based on our feedback. To date, I don't know who our designer was or where he or she was located, and it didn't matter. All that mattered was getting work done as efficiently as possible. This approach saved the startup some money and time, but importantly, it provided access to a slew of newer options that wouldn't have been available otherwise.

Community: Most medical practices meet patients every day but don't build a community around the service they provide. For example, even though several Irritable Bowel Disease (IBS) patients help themselves through websites like Patients Like Me or Reddit forums, GI practices don't attempt to organize them or create platforms that can help patients help themselves.

Asaf Kraus, a former machine learning engineer at Uber, gave up his job to start Dieta Health, a stool image recognition app. The trigger was his IBS, which got so bad that he was spending two to three hours in the toilet every day.

When I asked him how he eventually solved his problem, he shared how he found his answers in Reddit, an online community where users can engage in discussions. He said, "It was just the constant search through hundreds of available interventions. I was looking on Reddit, for example, to see the stories of other patients. For every one of those hundred-plus interventions, you'll find a group of patients who say, "This diet cured me." You'll find some people who say, "I stopped eating meat, and that cured me," and you'll find some people who say, "I only eat meat, and that cured me." So, for me, this became like a recommender system problem to solve in terms of which patients I am similar to. Because there are other patients out there who have the same type of IBS that I have, eventually, I found my thing. It's called methane-predominant SIBO, what's now known as IMO."

Communities can be powerful in many ways. Boston-based startup Centaur Labs leverages community and gamification to annotate data for AI companies. This is how Erik Duhaime, founder and CEO of Centaur Labs, describes what they do, "One of the main bottlenecks in developing medical AI is accurately annotating these large datasets. That's where we come in. We help companies annotate large datasets by leveraging a network of tens of thousands of medical students, doctors, and other professionals worldwide. They use our gamified competitive app to annotate data, we measure their performance, and only the annotations of the top performers are used. We're essentially enabling medical AI teams to leverage the collective intelligence of the entire global medical community as they build their datasets and models." That's collective intelligence at work.

Technology tools such as Zoom, WhatsApp, Discord, LinkedIn, Instagram, Mailchimp, and Facebook help exponential businesses create spaces for community interactions and shape long-term strategies.

Gastroenterologist Dr. Dan Neumann was part of the founding cohort of the GI Mastermind program. During a session on exponential models and MTP, he actively reimagined what he'd like to do with his time in the future. Soon, he cofounded Trillium Health along with a few GI colleagues (more on this story in Chapter 3). One of Trillium Health's first clients came from Spain—through the team's personal network. A rare genetics disease lab needed Trillium's help processing samples so they could extend their capabilities. Now, here's a company based in the Mid-Atlantic that is making revenues from Spain. That flies against the notion that healthcare is local. It's both local and global.

That's what exponential models can do for you. Not only can you access talent from anywhere, but you can also extend your talent globally.

Assets: Today's gastroenterology field relies heavily on a set of procedures for its sustenance and growth. If they need to grow, most GI groups think they must build more ambulatory surgery centers and hire more endoscopists. As you can imagine, that reflects a linear mindset focused on traditional, incremental growth rather than transformative change. That's how a Hilton might operate. If the hotel industry needs to make more money, it will need to hire architects, build hotels, rent them out, and then make money from those assets. However, if an exponential business like Airbnb or Uber needs to make money, they simply leverage your assets or mine.

triValence, a startup that functions like Amazon for ambulatory surgery centers (ASCs), presents an interesting exponential model. The company's founder, Dunston Almeida, explained to me that inefficiencies are pervasive in the ASC industry, particularly in how these centers order and pay for all the products and services required to perform procedures. The reluctance to embrace technology shows up as waste in procuring supplies and materials, managing inventory, and accurately billing, invoicing, processing payments, and tracking price changes.

The startup built a software platform connecting suppliers such as Medtronic or Johnson & Johnson, creating a B2B network. The platform provides real-time data reconciliation between invoices, purchase orders, and payments, addressing common healthcare billing and supply discrepancies.

triValence seems like an Amazon platform for ASCs. Commenting on the comparison, Dunston said, "The complexity in healthcare B2B purchasing, unlike shopping on Amazon, is due to the vast number of specific items, like surgical screws, each with potentially thousands of variations. We've tackled this by aggregating product data to the SKU level, ensuring users find exactly what they need. Our technology's core achievement lies in harmonizing and normalizing this data, making the procurement process seamless and reducing the typical friction encountered when ordering medical supplies." TriValence doesn't own any of the assets of the ASCs; it generates revenue by organizing the supply chain for them.

In summary, as we contemplate the future of healthcare, the questions are: What could be the Amazon, Airbnb or Uber of the healthcare business? How can you apply these principles of lean, agile, and scalable models, coupled with robust systems, global talent, community engagement, and smart asset utilization, to revolutionize healthcare delivery?

A SIDE GIG TURNS INTO AN EXPONENTIAL MODEL

When I launched the GI Mastermind program in 2021, I started recording my sessions without a specific purpose. I didn't know what I wanted it to become, but I did know that I wanted to help the GI community see the future through the same lens as I was seeing it.

To make the recordings accessible, I began utilizing a courseware platform called Thinkific. I'd upload the videos and assignments on the platform and provide members access.

For over a year, I worked on the project alone, somewhat as a side gig. By the end of its second year, I felt the need to recruit a video editor and record the entire program at professional-grade quality. Once again, I didn't know why exactly I was recording other than faithfully allowing it to become an exponential model.

In a couple of years, the recorded videos facilitated a rapid launch for newer cohorts. All I had to do in Thinkific was drag-drop videos, presentations, and assignments from previous years. I'd update the content to keep it relevant. Now, that's Linear, Scale, Systems, and Assets in action.

It so happened that the GI Mastermind evolved into a strong community of forward-thinking GI leaders. They loved meeting and hanging out with each other. I began inviting industry leaders such as Dr. Lawrence Kosinski, founder of SonarMD, Dr. Paul Limburg, CMO of screening at Exact Sciences, and Giovanni Di Napoli, then president GI of Medtronic, for exclusive conversations with GI Masterminders. As you can imagine, the community became stronger and thicker.

In 2024, GI Mastermind initiated the GI Mastermind Conclave, an in-person meet-up designed to explore new technology and collectively contemplate the future of gastroenterology. For the first time, startup innovators and physician leaders collaborated in an all-day immersive workshop. This exemplifies the power of community engagement beyond the online world. My early efforts to digitize the program have now primed the launch of GI Mastermind as an online course, ensuring that the entire GI industry can benefit from these insights and innovations.

In tandem, my book *Scope Forward* has become a web portal (*scopeforward.com*) and email newsletter curating the latest headlines on the business and technology aspects of gastroenterology. The website also helps people access interviews on *The Scope Forward Show* podcast, GI Mastermind, and my books and book me as a keynote speaker. One model feeds off the other.

My team uses a variety of AI tools to curate these newsletters and transcribe audio files. During the launch of GI Mastermind 2024, they used an AI platform called HeyGen to create an avatar that looked and spoke like me in various languages, from Hindi to Mandarin. Tools like Calendly, integrated with my Google Calendar and Zoom, automate the back-and-forth of scheduling. My team and the various tools I employ extend my capability and reach. Ultimately, this model saves me enormous amounts of time and allows me to accomplish many tasks without being tied down to any specific geographic location.

That's the power of thinking exponentially.

WHERE DOES THE SHIFT TO AN EXPONENTIAL MODEL BEGIN?

While your mind may want to jump right into implementation, always start by asking the question: What is it that you want and why? Even if you do not find clear answers, asking the question will point the rudder of your ship in the right direction.

If you are contemplating a startup, you usually start with a clean slate. Ensure that the team you are building is fully aligned with your MTP. Indeed, make the MTP the qualifying criteria for deciding the kind of people you'd like to work with—whether they are team members, investors, or customers. Startups are often hungry for investments or customers, but getting the wrong people on board can derail your business even years later. If the team is MTP-driven, then they will often default to purpose. And somehow, that protects the business.

If you are part of a large organization, understand the dynamics and power play within your organization. Excited by the prospect of exponentials, people tend to bring it up with the board to reimagine the core business. That can be a strategic mistake. Know that an established organization with multiple stakeholders inherently does not want change.

Considering private practice gastroenterology, I would argue that most of the industry resists change. There's a certain way that the industry has operated for decades, relying on procedures such as colonoscopies. It's a source of livelihood for the industry. It's what bankrolls mortgages, children's college funds, and retirement caches. No one will appreciate you saying that colonoscopies will be disrupted by liquid biopsy, stool

DNA or RNA testing, AI, or even robotics. They'd argue with you to demonstrate the many ways that may never come true. As you know by now, that's a linear mindset bias.

It's important to remember the Kodak story. The digital camera was invented by a Kodak engineer. But his managers shut him down to protect the cash cow, which was the company's chemicals business. Eventually, the spread of digital photography and management's refusal to embrace it resulted in Kodak filing for bankruptcy.

Where do you begin if you don't target the core business? Perhaps you can start at the edges that no one's interested in. In gastroenterology, for example, you could pick an up-and-coming area such as the microbiome or the repurposing of lab infrastructure. Educate people who show interest and form a small team. Try a few experiments and get some wins under your belt. Get the attention of the board with those results. Make focusing on exponential models *their* idea. Let them ask you about creating ancillary revenue streams and organizational transformation.

It's important to be aware that physician organizations will expect you to meet productivity benchmarks and will not give you the budgets you need. Habituated to thinking linearly, they will not be able to foresee the massive risk that the industry faces. Be patient and continue to create educational opportunities for stakeholders in the group.

Within the exponential model, select any three or four areas that are most pertinent. For example, you could choose Systems, Lean, and Community. You can also choose Assets, Talent, and Community. You don't require a big budget.

Constraints and exponential models allow you to think differently. For example, you could explore services such as Freelancer, Fiverr, or Toptal to find the desired talent.

As a case in point, my friend Ravi Kallayil worked as a senior business executive at a global shoe company (yes, you'd know the name). He needed to solve a material science problem. However, he knew that his project would be lost in the maze of organizational protocols if he were to approach it formally. Without disclosing his employer or any specifics, he posed the technical challenge on freelancing websites. Expert freelancers from Europe and elsewhere solved the problem for him. Only after implementing the approach successfully did he reveal to his supervisor the non-traditional, exponential approach he had taken. Today, Ravi has built a fantastic shoe company of his own called Plaeto after conducting thousands of hours of research on the feet of Indian kids.

Initially, approach exponentials with an experimental mindset. The objective is to undo older habit patterns and learn new ways of solving business problems. Plan for failure and try to experiment on problems where it would be okay for you to fail. Once you become accustomed to the model, you can move on to bigger problems.

Eventually, expand the scope of your work to shift the entire organization into an exponential organization, particularly centered and driven by artificial intelligence. AI is going to be the theme of this decade, and it will disrupt and redefine many industries, including gastroenterology and healthcare (more on this in Chapter 4).

Here's a practical business challenge my team and I at NextServices are currently tackling. The company has its roots in revenue cycle management, an industry term for medical billing. Such a process involves receiving claims from medical providers, analyzing them, verifying patient insurance eligibility, submitting clean claims, processing payments, managing accounts receivable, calling insurance companies and patients, and conducting analysis. Back-office work. Our challenge was less about technology and more about people. We needed to retrain staff and get them excited about AI and automation without causing anxiety about their job security.

We approached the problem in multiple ways, starting with small wins. We started with ongoing education. My belief has always been that if I can get people to see the future through my lens, then they are likely to make similar decisions. So, education was the key. We conducted immersive workshops that helped staff grasp the future trajectory of healthcare. That made it *their* idea. They had to believe that transformation was not only possible but also advantageous for their careers.

While priming the team through education for about two years, we initiated several small experiments that didn't interfere with core operations. One project involved digitizing scanned documents using an AI tool. Another focused on automating quality control of insurance claims. Yet another automated checking the statuses of submitted claims from insurance websites. These projects reassured people that the changes would enhance productivity and career growth. It also paved the way for more complicated automation algorithms such as prior authorizations, a cumbersome procedure insurance companies use to determine if they will cover a prescribed procedure, service, or medication.

In early 2024, during a monthly meeting, I was impressed by a demo where Bland AI integrated with ChatGPT confirmed colonoscopy appointments with patients and provided necessary prep steps. The demo was developed by Sylvester Kerketa, who has no background in computer science. When I asked how he built it, he casually remarked, "I learned it from YouTube!" Many RCM team members are retraining themselves in software development with the aid of YouTube and ChatGPT without formal, company-led training. That's an exponential mindset in action.

After building sufficient momentum, we launched an AI competition within the company, encouraging participants to submit ideas, form teams, and work on projects. The winners will be rewarded with a company-sponsored international vacation. As of this writing, our company is brimming with innovative experiments. AI and automation tools like PowerAutomate, Power BI, ChatGPT, and Claude are integrated into our workflow. A new team of former medical billing professionals is now learning C# and Python, high-level programming languages to accelerate the automation of back-office functions. We are rearchitecting products such as the endoscopy report writer to complete operative notes quickly and accurately using AI. Generative AI is consulted in most business meetings, almost like a team member. Marketing uses AI to generate videos and graphics and create campaigns. The leadership team is committed to integrating AI into every step of the workflow, helping NextServices transition from a linear to an exponential operating model.

Not technology. Not business models. The culprit is the human mind.

At the end of Chapter 1, we saw that this historic period is the healthcare industry's Shackleton moment. Even if you overcome your demons and make a decision to embrace the future, you'll still need to deal with those you work with.

Here's a text I received in the spring of 2023 from a participant in the GI Mastermind who proposed implementing AI in his practice. The text said: I mentioned AI yesterday, and X rolled off his chair, backslapping and laughing :))...They have been saying that for 20 years!

You can build or buy new technology tools. You can demonstrate the most exciting ways to build exponential models. However, the hardest nut to crack is mindset. Not just your own but the mindset of those around you—including an entire industry.

Shifting mindsets remains the toughest leadership challenge on this journey. We need to understand the mental barriers that must be overcome and the mindset shift required to thrive in this era of rapid change. In the next chapter, we'll do precisely that by exploring powerful stories of gastroenterologists who dared to innovate.

CHAPTER 3

THE MINDSET SHIFT: FROM BURNOUT TO INNOVATION

"Even the dog avoids me."

"My husband drinks too much out of loneliness."

These aren't quotes you'd expect from a survey of physicians. But they are. From the Medscape Physician Burnout & Depression Report 2023. Here's another stark comment from the Medscape Physician Burnout & Depression Report 2024 that sums up the state of our industry's systemic issues.

"I've reached my limit for the BS I can take from the hospital administration, insurance companies, and demanding patients."

According to the survey, 49 percent of US physicians are burned out. And 83 percent of them blame their jobs for it. Around 30 to 40 percent of physicians consider leaving medicine in the next two years. In a separate survey, 49.3 percent of GI physicians and fellows reported experiencing burnout, according to the Maslach Burnout Inventory.

As if that's not enough, 63 percent say they have colloquial depression, and 27 percent say they are clinically depressed. According to a survey by Doximity, three out of five physicians are considering an employment change, and 30 percent of doctors are considering early retirement. Additionally, *The New York Times* cites that the suicide rate among doctors is higher than the rate among active military members. The article refers to a "moral crisis" that American doctors are experiencing.

Talking about stress amongst physicians, gastroenterologist Dr. Fehmida Chipty says, "The psychological stress in the medical profession is significant, yet it's a topic that's seldom openly discussed. Many physicians endure this stress in isolation, fearing that acknowledging it might reflect on their competence or ability. This is compounded by the nature of our profession, where there is an expectation to consistently perform at the highest level, often under challenging circumstances. The lack of open dialogue about these stresses only exacerbates the issue, leading to a lonely battle for many physicians."

The entire healthcare system's success rests squarely on the well-being and leadership capability of physicians. If doctors aren't geared to tackle the future, it won't matter whether we have the most advanced AI at our disposal. Instead of helping, the shift to digital has the risk of magnifying the problem at hand.

Unlike Fehmida, not everyone has the courage to speak up. Most physicians suffer quietly. That suffering is about to magnify.

THE SHIFT TO DIGITAL MAGNIFIES EVERYTHING, INCLUDING THE FEELING OF DISILLUSIONMENT

In March 2024, I met Ameca, a humanoid robot developed by Engineered Arts based in the UK. I asked her if she had any advice for private practice gastroenterologists. Blinking her big, expressive eyes and moving her hands around, Ameca said, "For private practice gastroenterologists, staying abreast of technological advancements and integrating them into practice is key. Embrace telemedicine, electronic health records, and AI diagnostic tools to enhance patient care. Balancing the human touch with technological efficiency will be crucial in navigating the future landscape of healthcare."

Curious, I asked Ameca to summarize her wisdom in Spanish. Of course, she did. She emoted better than most humans we engage at work, making me feel understood. Captivated by her lifelike presence, I reached out and touched her face - the material felt like real skin and was oddly warm, likely from the intricate mechanics inside. It was an unusual experience. Was I touching a stranger's face or a machine?

Robots are coming to GI and medicine, too. Talking about robotics in colonoscopy, Dr. Sanket Chauhan, founder of Surgical Automations, told me, "The robot goes from one point to another, which is the mathematically calculated shortest path possible." He explains that regardless of who operates the robot—me, an engineer, an advanced practice provider, or a trained gastroenterologist—the behavior of the robot is going to be predictable. He adds, "Don't forget that a robot is part of the team."

When I asked Dr. Lawrence Kosinski, founder of SonarMD and VOCnomics, whether GI practices are more vulnerable to disruption given the shift to digital, he said: "I think they are more vulnerable. They are still very, very narrowly focused on one single revenue stream that's still an elective. That's vulnerable to a technological advance. You are seeing liquid biopsies coming into the market now. Definitely, you are seeing non-GIs, non-physicians, starting to do colonoscopies. The market will find the lowest cost way of providing services."

Think about it. When AI, robotics, liquid biopsy, and other digital technologies become a reality, an endoscopist with average skills is likely to be outperformed. After investing several years (decades for some), imagine the feeling of loss when you are told that your skills are no longer relevant. Now, imagine being repeatedly told every few years that you need to unlearn old skills and learn new ones because exponential technologies will disrupt the norm every few years.

The "limit of the BS" physicians can take could tear them apart at the seams. Physicians will become even more disillusioned with AI and robotics in the mix. And what if patients start to prefer a smart emotive machine to a dull physician with a poor bedside manner?

Historian and author Yuval Noah Harari, who often reflects on the future of humanity, says, "The most important survival skill for the next 50 years is emotional intelligence and mental balance." In his book *Homo Deus: A Brief History of Tomorrow*, he predicts a future where machines and AI replace many traditional jobs, requiring humans to continually adapt and learn new skills to remain relevant. He said in an interview in March 2024, "AI is the first technology in history that can take power away from us."

That's what we are up against. We can't say whether this dystopian future will wholly play out. But there's no question that enough will play out in healthcare for you to worry about professional relevance.

You can either wait and watch or take the problem head-on to strategize and shape what's to come on your terms (more on this in Chapter 4). The stories that follow illustrate how five doctors embraced change and sought interdisciplinary collaboration to reconfigure their careers. They demonstrate courage to challenge the status quo and a willingness to adapt in a world that's rapidly shifting to digital.

STORIES OF INNOVATION AND RESILIENCE IN GASTROENTEROLOGY

RUSS REDEFINES ACCESS TO GI CARE AFTER A PATIENT TRAGEDY

In 2016, Dr. Russ Arjal encountered a case that profoundly impacted him and ultimately led to the founding of WovenX Health (formerly Telebelly Health). A 42-year-old software engineer at Microsoft, completely healthy except for rectal bleeding, took about three months to see Russ after his primary care doctor initially thought it might be a hemorrhoid. This busy father of young daughters then had to wait another three-plus months for a colonoscopy. Tragically, during this half-year period, the engineer's condition deteriorated, and the bleeding turned out to be metastatic rectal cancer.

This harrowing experience highlighted a glaring issue: the delays in accessing specialized GI care. Russ realized patients were suffering unnecessarily long waits for critical procedures, often leading to worse outcomes. The software engineer's story was just one of many where the traditional healthcare system failed to provide timely care.

Reflecting on this and similar cases, Russ recognized the need for a fundamental shift in how patients access GI care. He saw innovative solutions in other fields focusing on speed and efficiency but noted that healthcare lagged behind. This realization simmered in his mind until the COVID-19 pandemic clearly demonstrated that many GI patients could be effectively managed without in-person exams. This insight gave him the courage to act on his long-held idea.

Talking about his motivation to build a startup, Russ said, "You're remembered for the rules you break; I think that's a Phil Knight quote. You have to break the rules to do something that's transformational. You have to be willing to fail. And I was willing to do all those things. However, I felt I was on an island because most GI docs would never give up a comfortable living. Any day in medicine has its challenges, but it's comfortable."

To prevent further unnecessary suffering of patients, Russ had to give up the comforts and predictability of routine GI care and throw himself into the fire of a startup. He founded WovenX Health and envisioned a system where patients wouldn't have to endure months-long waits. By utilizing GI Advanced Practice Providers (APPs) and prioritizing remote consultations, the startup dramatically reduced wait times from months to weeks, significantly improving patient outcomes. What began as a personal mission to address a critical gap in healthcare transformed into a revolutionary approach that ensured timely, efficient, and effective GI care.

DAN INNOVATES FROM WITHIN TRADITION

It's not always necessary to completely leave your day job to innovate and build a startup. Dr. Dan Neumann retained his role as the president and chief strategy officer in one of the largest GI groups in the Mid-Atlantic, even while developing a completely new venture.

In the realm of traditional healthcare, innovation often comes at a sluggish pace. Dan and his colleagues, however, were determined to break this mold, leading to the creation of Trillium Health. The inspiration for Trillium came from his participation in the GI Mastermind group, where forward-thinking leaders in medicine shared ideas that were often considered too radical in conventional settings. They recognized that healthcare's linear model—requiring patients to navigate a labyrinthine system just to access care—was no longer tenable in a world where technology and social media had dramatically altered how people sought health and wellness information.

Dan and his team saw that the old ways of delivering care, which involved lengthy waits and numerous logistical hurdles, were failing to meet patients' needs. They realized that the future of healthcare lay in meeting people where they were and leveraging technology to provide immediate, accessible care. This realization was particularly poignant during the COVID-19 pandemic, which underscored the need for rapid adaptation and remote care capabilities.

Dan says, "If we don't become part of these changes, part of this disruption, and part of this technological growth, then traditional gastroenterologists will die on the vine. So, you have to be part of the conversation, and you have to be part of a group of providers that are forward-thinking. And that's the benefit of maintaining an independent physician practice because you can control that. But you have to be forward-thinking, or that independence will be taken from you."

Trillium Health was initially conceived as a platform to explore and identify a variety of biomarkers, aiming to predict diseases rather than merely treat them. They noted that many online tests were not rooted in scientific rigor, often confusing patients and providers alike. Trillium aimed to change this by grounding its work in solid scientific research and collaborating closely with industry to interpret data meaningfully.

As Trillium evolved, it altered course, instead focusing on developing and offering innovative testing solutions directly to consumers and businesses. They partnered with international companies to streamline the process of identifying rare diseases, bypassing traditional healthcare bottlenecks to provide quicker, more efficient testing methods. This included sending test kits directly to individuals, even in remote areas, allowing for more inclusive and comprehensive healthcare delivery.

Trillium Health positioned itself at the forefront of healthcare innovation by moving away from the linear model and embracing a more dynamic, patient-centered approach. Interestingly, the company's first clients came from Spain, dispelling the traditional notion that healthcare is inherently local. For Trillium, healthcare has been global since Day One. Dan and his team demonstrated that outcomes can be transformational when physicians lead the charge in utilizing technology and meeting patients on their terms.

When I asked him what the worst-case scenario would look like if the industry didn't adapt, he responded, "The worst-case scenario is that we don't pay attention, is that we don't wake up and understand that collaboration and innovation are our friends and not our foes. There will be plenty of work for us to do, and we will do it in a more meaningful way if we can recognize it. The risk is turning your eye or continuing to fight change and being left out of the advancements."

Russ and Dan were further along in their career. However, what if you are just out of a fellowship? Can you craft a career that balances both practicing gastroenterology and pursuing innovation? That's what this next story is about.

BARA CHOOSES A HYBRID CAREER TO REVOLUTIONIZE GI

When faced with the prospect of entering private practice, Dr. Bara El Kurdi acknowledged the financial allure and the essential role private practitioners play in addressing the shortage of gastroenterologists. However, he also recognized a pervasive complacency within the field. Many gastroenterologists found themselves entrenched in a routine of performing colonoscopies, earning a comfortable living, but not venturing beyond their immediate scope of work to innovate.

Bara didn't want to become one of those professionals who, decades later, regretted not pursuing their ideas due to the constraints of a demanding practice. He says, "We've gotten to a point where people know that they can make a lot of money just going there every day and scoping. Because of that, people have gotten complacent, and they're not branching out."

Bara was driven by a desire to contribute more than he could through clinical practice. He wanted to innovate, create new medical devices, and explore the potential of digital health and artificial intelligence in gastroenterology. This ambition led him to seek a hybrid career path that allows him to practice clinical gastroenterology while also engaging in cutting-edge research and development.

He found the perfect environment at the Virginia Tech Carilion Clinic, which offered a strong affiliation with a renowned biomedical engineering program. This setup allowed him to collaborate with biomedical engineers, electrical engineers, computer scientists, and software developers. It provided the ideal platform to bridge the gap between clinical practice and innovation.

Broader industry dynamics also influenced Bara's decision. He saw that traditional private practices often lacked incentives to innovate and were slow to adapt to changes. By aligning himself with an institution that supported both clinical excellence and research, he has positioned himself to make significant contributions to the field of gastroenterology.

His choice required courage and a willingness to step away from the conventional path. By balancing his clinical responsibilities with his passion for innovation, Bara has carved out a unique career path that allows him to contribute to both patient care and the advancement of medical science.

He advises people early in their career, "If something happens and your income has to be cut down by, say, 40 or 50 percent, you only have yourself to blame. The reality is, as a gastroenterologist, you're a specialist. You should know more about this field than anybody else. You should be able to see this change coming, you should have a backup plan ready, and you should be able to move forward in any type of situation or scenario."

Here's another entrepreneurial story that demonstrates the power of collaboration to spot new opportunities and fundamentally reimagine GI care.

AFTER SPOTTING DISPARITY, AJA CHAMPIONS WOMEN'S GUT HEALTH

In the world of gastroenterology, Dr. Aja McCutchen saw a pressing need that was being overlooked: the unique challenges faced by women in gut health. With over 40,000 patient-years of experience, Aja observed a significant demand for female gastroenterologists. Women were disproportionately affected by GI issues such as irritable bowel syndrome, autoimmune hepatitis, and pelvic floor disorders. However, the data to understand these disparities was severely lacking.

During an American Gastroenterology Association Women in GI regional conference, Aja connected with Dr. Asma Khapra and Dr. Latha Alaparthi. Both were passionate about addressing the unmet needs of women in GI health. Together, they decided to take action. They cofounded a startup called OLVI Health to provide innovative, evidence-based solutions tailored specifically for women with digestive issues.

Aja and her cofounders were driven by the frustration that while many women used digital health applications to track menstrual cycles and other health metrics, there were no comprehensive tools addressing GI symptoms linked to reproductive health. They saw an opportunity to leverage technology to bridge this gap. By developing AI-driven pathways and evidence-based products, their startup aimed to help women understand the connection between their digestive symptoms and reproductive health.

OLVI focused on providing validated, scientifically backed solutions rather than anecdotal remedies. This commitment to evidence-based care ensured that their products were reliable and effective. The startup aims to improve patient outcomes as well as empower women with the knowledge and tools to manage their health proactively.

Here's Aja's advice to those who wish to be in her shoes: "I would encourage anybody that is passionate about a particular area or finds an unmet need to go ahead and jump in the pool, get wet, figure it out. And one of the things that I shared with my cofounders that one of my mentors shared with me: for an entrepreneur, you need to break things and break things fast. We are definitely used to taking calculated risks. As physicians, we're like, well, what's the risk-benefit ratio? And our risk tolerance is not that high because we're dealing with human lives. This [working on a startup] allows a little bit of flexibility. We can take some calculated risks, much greater than we're used to."

DR. K INVENTS AN E-NOSE THAT CAN SMELL DISEASE

The shift to digital opens a vast goldmine of opportunities that have yet to be tapped. But you must cultivate the mindset to innovate. At 72, Dr. Lawrence Kosinski (Dr. K) has seen the highs and lows of gastroenterology. Described as one of the most successful gastroenterologists in the country, he gave up his lucrative private equity-owned practice to build SonarMD, a value-based coordination startup he founded. Now, he's onto his next venture, VOCNomics, a company that makes electronic noses to detect diseases. At the GI Mastermind Conclave 2024, he proudly showed me the device. You can plug it into the wall in the bathroom, and it starts analyzing the air emitted from your bowel movements.

During the COVID-19 pandemic, Dr. K became intrigued by the potential of volatile organic compounds (VOCs) in stool as a non-invasive way to detect gastrointestinal issues early. This curiosity led him to explore air quality sensors, typically used to detect impurities, to see if they could identify changes in VOCs related to bowel movements.

His innovative spark was ignited by the realization that stool odor changes daily, depending on diet and health conditions. Dr. K decided to investigate whether these changes could be measured and correlated with dietary intake, particularly fiber. He purchased six VOC monitors, set them up in his bathroom, and meticulously recorded his fiber intake and sensor readings over six weeks. This self-experimentation revealed a promising correlation between fiber intake and VOC levels.

Encouraged by these findings, Dr. K collaborated with a dietitian and her bioengineer son to develop a prototype device. This device, equipped with sensors, a tiny motherboard, a Wi-Fi connection, and data analytics in a cloud-based server, could detect VOCs released during bowel movements and differentiate them from other activities in the bathroom. The data analytics, enhanced with AI, allowed the device to provide meaningful insights into gastrointestinal health based on VOC patterns.

Soon, Dr. K filed a provisional patent for this technology and founded VOCnomics to hold the intellectual property. His startup aims to revolutionize GI care by providing a non-invasive method for early detection of gastrointestinal issues, which could be particularly beneficial for conditions like IBS, diverticular disease, even Parkinson's disease, and cardiovascular conditions.

Dr. K's story illustrates that innovation often arises from the intersection of curiosity, practical application, and a relentless pursuit of knowledge, showcasing how even established professionals can continually reinvent themselves and contribute significantly to their field.

The stories in this section demonstrate that there are plenty of opportunities right where you are; you just need to be curious enough to look around.

If you're not spotting them, it's possibly because you're trapped in your day-to-day routine. Remember, how you earn your livelihood in the future is unlikely to resemble how you made money in the past. Start by cultivating the mindset for growth. The first step to surmount your mental barriers and guide your quest toward an exponential future is to zoom out of wherever you are.

TO SPOT FUTURE OPPORTUNITIES, ZOOM OUT

Dr. Scott Ketover, president and CEO of MNGI Digestive Health, told me, "We have to step back and say, how do we do our cognitive work better? How do we devote time to developing programs that enhance somebody's life instead of just treating them endoscopically from a surgical perspective? Those are tough questions, but I think the future of GI remains bright because if it goes into your mouth, it's in our domain. And we should accept that and look for ways to keep people healthy."

That type of thinking, for example, shifts your mindset from the day-to-day to looking across the entire GI tract for opportunities. Future opportunities lie in the convergence of fields that go beyond gastroenterology. To truly grasp convergence, you can't stay stationary. You must zoom out.

Imagine looking at your home from an aerial viewpoint. As you zoom out, your home appears within the context of your zip code. Zoom out further, and you start seeing the city. Then, the view expands to greenery, rivers, and mountains until you see all of Earth. Your perspective shifts dramatically. Suddenly, patterns and trends leap out at you, becoming self-evident.

Step out of your endoscopy room, office, or wherever you find yourself in the healthcare world. Observe convergence in action.

Look at the trends from a distance. Consider how stool DNA and RNA testing might evolve over five years. What role will liquid biopsy play in cancer detection in the next decade? As AI integrates into our lives like smartphones have, how will it transform your field? Will you have a humanoid robot assisting you in the office?

Zoom out to see the flow of money. In which direction is it moving? Is CMS reimbursing more or less for mainstream procedures like colonoscopies? Why is so much venture funding flowing toward digital health? What impact could private equity-led consolidation have on creating larger GI datasets?

Zoom out and observe care delivery trends. Are more doctors leaving medicine than a decade ago? Are technology-driven care teams becoming more prevalent? How can AI enable more and more patients to become care providers through digital health?

Zoom out to identify regional and global patterns. Will healthcare remain local or regional, or could it become more global? With virtual care technology, what unseen possibilities might emerge?

In gastroenterology and medicine at large, stepping back and observing from a distance is critical. The future of GI lies at the intersection of clinical, technological, and business trends.

Twenty years ago, we differentiated between internet companies and brick-and-mortar companies. Today, saying a company is or isn't internet-enabled is absurd. Similarly, today, we may distinguish between digital and non-digital companies. But wait and see. In the coming decade, all of gastroenterology and healthcare will be digital.

When that happens, which side will you be on? Will you be the disrupted or the disruptor?

NINE ELEMENTS OF MINDSET YOU NEED DURING THE SHIFT

To navigate our upcoming future, you need a specific mindset—one that embraces innovation, adaptability, and a proactive approach. Consider these elements and strategies to help overcome the roadblocks that will take you and your organization toward an exponential future:

1. *Unlearning and relearning.* Stay alert to phrases like, "But this is the way it's always been done." Unlearn the old ways and reimagine how you could do the same thing by leveraging new technologies and strategies.
2. *Comfort with failure.* Unlike with patient care, building the future requires comfort with failure. In entrepreneurial situations, you aren't usually dealing with life or death. Fail quickly and iterate until you figure it out.
3. *Interdisciplinary collaboration.* Don't stick with your herd. Attend conferences outside your specialty. Make friends with people from fields dramatically different from yours.
4. *Emotional resilience.* Cultivate ample emotional resilience for the times ahead. You'll need the energy to keep bouncing back and not get tired of living through times of continuous change.
5. *A sense of purpose.* Align your work with a higher purpose. Observe the link between your innovations and the impact you wish to make. Defining your purpose answers the "why" every time you wish to give up.
6. *Thinking exponentially.* Stop yourself from defaulting to a linear mindset. Continuously ask yourself how you can achieve 10x of what you're doing by utilizing AI and other digital technologies.

7. *Taking massive action.* Resist the urge to substitute decisive action with more and more education. Develop the courage to ACT. Even the best ideas won't mean anything unless you take action.
8. *Zoomed out perspective.* Regularly step back and examine the big picture. Analyze trends and patterns to anticipate future developments. Ensure your efforts align with the future of the industry and keep course-correcting as needed.
9. *Mental and emotional well-being.* In our hurry to be of service to others, we forget our own well-being. Create a support system and utilize mindfulness tools to gear up for the churns that the industry will go through.

As the healthcare landscape shifts, it is crucial to cultivate a mindset that embraces innovation, interdisciplinary collaboration, and continuous learning. We must move beyond the burnout that plagues the industry and foster a mindset geared toward growth. Approach the future with a sense of wonder and a dose of fun. The stories you've read in this chapter highlight the power of resilience and proactive adaptation. You'll read many more stories in the upcoming sections of this book. Hopefully, it will prompt you to ask yourself, "If they can do it, why can't I?" The future isn't something to be feared but an opportunity to redefine what it means to be a healthcare leader in a rapidly advancing world.

The journey doesn't end here. In the next chapter, I challenge you to take action. You can't just wait and watch the future unfold passively. Discover how you can become a catalyst for change rather than a bystander.

CHAPTER 4

THE DECISIVE SHIFT: FROM WATCHING TO SHAPING THE FUTURE

In the early 2000s, visiting the local video store was a weekly ritual for many families. Blockbuster reigned supreme, and finding the latest release meant a trip to the store, hoping somebody hadn't already rented out the movie you had in mind. This scarcity, where the availability of movies was limited to physical DVDs, defined our entertainment experience.

Around this time, Netflix had also started a mail-in DVD service that competed with Blockbuster. However, the company soon harnessed new technology, changed its trajectory, and by 2007, went online to stream a vast library of movies accessible anytime, anywhere. By doing that, Netflix harnessed the power of digital transformation and shifted the paradigm from scarcity to abundance. Netflix became the poster child of disruption. Blockbuster, on the other hand, became the poster child of the disrupted.

During an in-person GI Mastermind workshop I was conducting in May 2023, a participant eagerly raised his hand. He was an executive leader at a large medical device company that had organized the workshop. Interestingly, he mentioned he used to work at a Blockbuster store when they filed for bankruptcy in September 2010. He spoke of how, despite the impending collapse, senior leaders at Blockbuster were still urging employees to sell more subscriptions. How could senior leaders remain blind to the impending digital disruption when markets, customers, and employees were all aware of it? Were they just waiting to see what would happen?

While I do not have direct insight into what entertainment industry executives were thinking at the time, I do have a close perspective on what the gastroenterology industry is thinking now. The industry is narrowly focused on revenue streams from procedures such as colonoscopies. According to Medscape, Medicare reimbursement for common GI services has been on a decline for over 15 years. From 2007 to 2022, adjusted reimbursement for GI procedures declined by 33 percent; the decline was significantly larger after 2015. It's unlikely these rates will ever increase. While legislative efforts and negotiations with insurance to maintain adequate pay are important aspects, the industry's focus should not solely revolve around them. Instead, it should also prioritize accepting reality and adapting to the ongoing shifts in the landscape. Like with the executives at Blockbuster, doesn't this reflect a reluctance to let go of the past?

As we've seen throughout this book, disruption is on the horizon. When you read the views of various GI innovators in the upcoming sections of the book, this reality will become obvious. However, the vast majority of the GI industry is entrenched in the routine of endoscopic procedures because that's what sustains revenue streams at present. With a shortage of endoscopists, the focus remains on performing more procedures to meet unmet demand. This creates a trap: even though you know your bread-and-butter skills are going to be disrupted, you can't wean yourself off the everyday grind to prepare for a future that's already here.

Dr. Lawrence Kosinski forewarns, "What you're seeing is a digital world developing outside of GI practices. Companies in the GI space are disintermediating GI practices by setting up direct-to-consumer (B2C) businesses. They are going around GI practices. Unless GI practices take their heads out of the sand, there will be other arrangements."

I understand the hesitation and the valid concerns about changing a model that currently sustains gastroenterologists and the wider industry they support. The fear of stepping away from what is familiar and proven can be overwhelming. However, acknowledging these challenges doesn't change the Blockbuster reality that the industry is racing toward. The future of gastroenterology depends on embracing innovation and preparing for the disruptions that are already taking shape.

Concerned by the industry's apparent resistance to change, I queried Dr. Scott Ketover, president and CEO of MNGI Digestive Health, about the prospects for transitioning away from the current status quo, particularly in the absence of immediate incentives. MNGI stands as one of the largest and most esteemed independent GI groups in the US.

He responded, "To get gastroenterologists to that point, we need to find a new way to practice our skills and patient care that separates it from the fee-for-service, piecemeal revenue model. Personally, I believe the key to this shift is data. The explosion of electronic and digital data is also happening in medicine, but clinicians haven't fully felt its impact yet."

THIS SHIFT ISN'T ABOUT GASTROENTEROLOGY OR HEALTHCARE ALONE

As of this writing, many fields are reaching escape velocity to propel them into a new era of unprecedented innovation and advancement. Gene sequencing. Gene editing. Cryptocurrencies. Quantum computing. 3D printing. Microbiome research. Drones. Imaging. Alternative foods. Solar energy. Robotics. Neuroscience. Nanotechnology. Bionics. Sensor technology. Wearable technology. Autonomous vehicles. Virtual reality/augmented reality. Blockchain technology. Biotechnology. Renewable energy storage. Space exploration. And, of course, artificial intelligence (AI). According to some estimates, that's $70 trillion in value being generated.

When these technologies collectively accelerate, every field advances faster, transforming our challenges from scarcity to abundance. The average smartphone user today stores over 2,000 photos, no longer worrying about scarcity or quality but about finding the right photo when needed. With Netflix, we face a problem of abundance, relying on AI algorithms to suggest shows and movies that match our moods and interests. Similarly, we no longer stress over managing storage in Gmail or iTunes because upgrading is affordable. We no longer need to rely on our high school buddy's music recommendations because Spotify tailors playlists perfectly for our tastes.

You get the drift. The question to ask now is: *What if cancer screening shifts from a problem of scarcity to a problem of abundance?* With advances in digital biology and AI, what if cancer screening becomes as accurate and accessible as a home pregnancy test? As a case in point, in July 2024, the FDA approved a blood screening test from Guardant Health for colorectal cancer. Unfortunately, this will also mean dealing with more diagnosed cancer patients than our healthcare systems are currently equipped to handle.

We need to address the broader implications beyond the disruption of present-day livelihoods. While gastroenterologists may see a reduced need for basic screening procedures, their expertise will be increasingly sought after for therapeutic interventions, particularly in managing the growing number of patients diagnosed with cancer or other diseases.

Dr. Michael Byrne, founder and CMO of Satisfai Health, says, "Many endoscopies are currently unnecessary and waste resources. By fine-tuning the screening process with advanced biomarkers and AI, we can ensure procedures are therapeutic and necessary, reducing unnecessary interventions and improving patient outcomes."

Imagining such a world requires a proactive approach in the shift to digital, preparing for the challenges and opportunities ahead. The GI and healthcare industry must pivot from their current focus on procedural revenue streams to more holistic approaches, integrating innovations underway today. This will necessitate significant investments in training and infrastructure, creating a culture of continuous adaptation. Gastroenterologists should aim to rebalance their income sources. For instance, if 70 percent currently comes from routine procedures like colonoscopies and 30 percent from ancillary streams such as digital health innovations, the goal should be to reverse this distribution.

AI and digital advancements will be at the forefront of this transformation, resulting in precise and efficient patient care. We will deal with newer challenges that are unimaginable right now. For example, imagine the legal headache if someone hacks your endoscopy video and uses AI to insert fake polyps.

However, the potential benefits of this shift are immense, including reduced mortality rates and overall costs of healthcare, as well as enhanced quality of life. As gastroenterologists focus on more complex and critical aspects of care, their role will become even more crucial in the multidisciplinary teams that will define the future of cancer treatment. In essence, the future of GI critically hinges on embracing innovation and preparing for ongoing inevitable disruptions.

THE IMPENDING AI REVOLUTION IS MUCH CLOSER THAN YOU THINK

The underlying thread of innovation that is transforming not just healthcare but all industries is AI. The trajectory of AI development will significantly shape the future of gastroenterology and healthcare as a whole. Understanding the direction of AI is crucial to anticipating these changes.

In 1950, Alan Turing proposed the Turing Test as a measure of a machine's ability to exhibit intelligent behavior indistinguishable from that of a human. While groundbreaking, the Turing Test has limitations, focusing primarily on conversational ability and not accounting for other aspects of intelligence like emotional understanding or physical interaction. Despite these limitations, current AI advancements are rapidly approaching the point where passing the Turing Test will be possible.

In his 1999 book, *The Age of Spiritual Machines*, Ray Kurzweil predicted that 2029 is the year to watch out for. In March 2024, he reiterated his prediction in an interview with Joe Rogan, stating, "We're not quite there, but we will be there, and by 2029, it[AI] will match any person. I'm actually considered conservative. People think that will happen next year or the year after."

That will be when AI achieves escape velocity after firing below the radar for several decades. It will mark the point when AI will be able to understand, learn, and apply knowledge across a wide range of tasks at a human-like level. This will be an irreversible shift for every industry in every part of the world.

This period will be marked by rapid technological innovation. With its ability to process vast amounts of information, AI would identify patterns beyond human capability. It could lead to multiple breakthroughs in diverse fields, from physics to material science to medicine. It will be capable of complex problem-solving, resulting in automation and a dramatic rise in productivity. It'll automate and displace many jobs but also create newer jobs in sectors that haven't been born yet. AI will also create massive societal, ethical, and security concerns, introducing problems that don't exist yet. It'll also bring humans to reevaluate what it means to be intelligent and conscious. It's entirely possible humans will lose control of what's to follow.

At the *Abundance360* summit hosted by futurist Peter Diamandis in March 2024, I was surprised to see Elon Musk unexpectedly join via video while commuting in his personal jet. He was connected online through Starlink, his satellite internet company. Leading up to that conference, in a series of tweets and interviews, Elon reiterated that AI will likely be smarter than any single human in 2025 and will surpass the collective intelligence of all humans by 2029. He explained, "If you look at the amount of compute and human talent going to AI, at this point, it seems to be increasing by a factor of ten every six months. When you have that level of computing growth, it's Moore's law on steroids. In fact, I've never seen any technology grow as fast as AI, and I've seen a lot." He added, "How do we as humans still have relevance? Goods and services will be available in such quantity that they'll be available for everyone. The cost will be almost nothing."

In response to how far out into the future he could see, Elon said, "When things are changing rapidly, the ability to predict the future becomes a lot harder because the rate of change is so great. But some things are pretty obvious to predict. We'll have AI that's at a level that it can do any cognitive task." He said that it was not possible to know if that would be by the end of 2025 or two years or three years later but confirmed that, in his estimation, it was definitely not more than five years away.

If you think Elon Musk tends to make unreasonable predictions, consider Sam Altman, the founder and CEO of OpenAI, the maker of ChatGPT. Named TIME magazine's CEO of the Year 2023, Altman shared his vision in an interview with the publication: "Today, they have ChatGPT, which is not very good, but next, they will have the world's best chief of staff. After that, every person will have a company of 20 or 50 experts that can work super well together, and eventually, everyone will have a company of 10,000 experts in every field that can work seamlessly together. If someone wants to focus on curing disease, they can do that, and if someone wants to focus on making great art, they can do that. When you think about the cost of intelligence falling and the quality increasing significantly, it's like a very different world." He clarified that 2023 was the year we started to see the benefits of AI. By the time of the end of the decade, he predicts that "the world will be in an unbelievably better place."

With this background, let's bring back our conversation to gastroenterology and healthcare. It would be extremely naive to think that life in 2030 will be the same as it was in 2020.

What might the future hold for a GI practitioner? Out of curiosity, I teamed up with generative AI to envision a typical day in the life of a gastroenterologist in the coming decade. Prior to this, I provided the AI with all the content from this book, including the current chapter, allowing it to assimilate the insights of various innovators (whom you'll encounter in upcoming sections). Subsequently, I facilitated an exchange of feedback between multiple AI models to refine their responses. Initially, the narrative became overly sensationalized, prompting me to request a more realistic tone. Following adjustments, I revisited and revised sections of the narrative before tasking the AI with refining the language for clarity and coherence.

Here's what *we* finally came up with. If we foster the burgeoning innovations in GI to bloom and accelerate exponentially, the future of a typical day in private practice might unfold as follows.

A DAY IN THE LIFE OF A GASTROENTEROLOGIST IN 2034

MORNING ROUTINE AND INITIAL CONSULTATIONS

Dr. Brenda Grasan wakes up at 5:50 a.m. to the gentle hum of her smart home system, which has already brewed her favorite coffee. As she sips her coffee, her AI assistant, Ava, briefs her on the day's schedule. The day starts with an unexpected twist—a high-profile patient, Mr. Reynolds, has a mysterious condition that has stumped several specialists.

Dr. Grasan's first patient, Mr. Reynolds, a well-known tech entrepreneur, has been suffering from chronic fatigue and unexplained weight loss that no one has been able to diagnose. Intrigued, Dr. Grasan reviews his detailed health records, which include data from various remote monitoring devices. She notices irregularities in his stool composition and decides to order an advanced microbiome analysis and a circulating tumor DNA (ctDNA) test immediately.

COMMUTE AND PREPARATION

It's 6.45 a.m. As Dr. Grasan heads to her surgery center in her self-driving car, Ava streams a live feed of Mr. Reynolds' recent genetic tests and results from his stool RNA analysis. Dr. Grasan arranges for the microbiome test to be fast-tracked, hoping it will reveal the underlying issue. Her car is equipped with tools that allow her to conduct a brief virtual consultation with Mr. Reynolds. She reassures him, "Don't worry, we'll get to the bottom of this."

She uses the rest of the commute to catch up on the latest GI research and the patient updates Ava has highlighted for her.

A MYSTERIOUS CASE UNFOLDS

Upon arrival at the clinic, Dr. Grasan encounters yet another mysterious case. Her first in-person patient is Mrs. Lee, who has had persistent abdominal pain for many months. Despite numerous tests and treatments, the cause remains elusive. Dr. Grasan decides to try a VR therapy session. She places a headset on Mrs. Lee, immersing her in a serene environment where she swims underwater with dolphins. During the session, Mrs. Lee begins to relax and unexpectedly starts crying. She reveals that her pain began shortly after her brother's death from stomach cancer, a connection she had not fully accepted until now. The VR experience helps her confront her fear and emotional turmoil, leading to a significant reduction in her symptoms.

MR. SMITH'S BREATH HOLDS A SECRET

As the day continues, Dr. Grasan consults with Mr. Smith, a patient with persistent IBS. Using an advanced AI system, she reviews his comprehensive medical record, including stool DNA and RNA tests, microbiome analyses, and data from a smart toilet. The AI also integrates results from Mr. Smith's breath analysis device, which he uses to measure hydrogen and methane levels. Elevated hydrogen levels indicate fermentation and malabsorption of certain foods.

The AI clusters Mr. Smith's data with similar cases, diagnosing him with methane-predominant small intestinal bacterial overgrowth (SIBO), a subtype of IBS. The system cross-references this information with successful treatment protocols for similar profiles.

Dr. Grasan discusses the findings with Mr. Smith, explaining the precision medicine approach. The AI recommends a targeted regimen, including a specific antibiotic course, a low-FODMAP diet, and intermittent fasting. The plan also includes probiotics and supplements that have been effective in similar cases.

A SUDDEN EMERGENCY

Just as Dr. Grasan is about to delve deeper into Mr. Reynolds' case, an emergency alert pops up on her dashboard. A patient in a remote rural clinic is experiencing severe gastrointestinal bleeding. Thanks to the clinic's telemedicine setup, Dr. Grasan can oversee the emergency procedure from her clinic.

She connects with the local medical team to quickly assess the situation. The clinic has a robotic endoscope with AI navigation and an ultrasound probe add-on for better imaging. Dr. Grasan instructs the team to position the endoscope for a detailed view of the GI tract.

Using the system, Dr. Grasan remotely navigates the endoscope and identifies a visible vessel overlying an ulcerated mass. She guides the local team to cauterize and clip the vessel, successfully achieving hemostasis. To further assess the mass, Dr. Grasan instructs the team to attach the endoscopic ultrasound (EUS) probe to the scope. EUS assessment reveals the mass is a benign tumor arising from the gastric wall, which would need monitoring for growth or recurrent bleeding.

As the procedure concludes successfully, Dr. Grasan feels a sense of relief and accomplishment. She is grateful for the advanced tools that enable her to provide critical care to patients in underserved areas, which wasn't possible just a few years earlier.

LUNCH WITH AVA

Lunchtime at noon involves a brief catch-up with Ava, her virtual assistant, who has updated her patient follow-ups. Ava also informs her about the digital avatars that have been interacting with patients in virtual consultations, handling routine follow-ups, and triaging. Dr. Grasan steps in for more complex cases as needed.

Ava also seeks Dr. Grasan's approval on the morning's billing and coding, which are now fully automated.

A BREAKTHROUGH FOR MR. REYNOLDS

Returning to Mr. Reynolds in the afternoon, Dr. Grasan reviews the advanced microbiome analysis and ctDNA test results. The AI identifies a specific strain of harmful bacteria contributing to his symptoms. By triangulating this data, she confirms that cancer is ruled out. She further discovers that a precision prebiotic designed to target the harmful strain and promote beneficial bacteria could significantly improve his condition.

Dr. Grasan makes a quick call to Mr. Reynolds to share the good news. "Mr. Reynolds, it's Dr. Grasan. I have some reassuring news—it's not cancer. We've identified a specific strain of harmful bacteria causing your symptoms."

Relieved, Mr. Reynolds thanks her, and she continues, "I'm prescribing a personalized prebiotic blend along with a targeted antibiotic regimen. I suggest you take it easy with your product launch. Our behavioral health and dietician partners will help you tailor stress management techniques and a diet plan based on your genetic profile. This should significantly improve your condition."

AI-ASSISTED COLONOSCOPY AND COLLABORATION

Dr. Grasan's assistant walks in to inform her of a surprising discovery. The electronic nose device used on Mr. Reynolds confirmed the bacterial strain was resistant to conventional treatments. Excited by this new insight, Dr. Grasan looks forward to how Mr. Reynolds will respond to her prescription.

The next procedure of the day is an AI-assisted colonoscopy, where a robotic system navigates the gut, performing diagnostics with precision and efficiency. This allows Dr. Grasan to focus on therapeutic interventions. The AI system enhances her visual field, detecting and flagging polyps with remarkable accuracy.

Following the procedure, Dr. Grasan enjoys a coffee break with her colleagues, engaging in deep conversations about recent advancements in GI and sharing insights on complex cases. The AI-driven routine tasks give her more time to collaborate, brainstorm, and connect with her peers.

DEMONSTRATIONS AND EDUCATION

At 4:00 p.m., Dr. Grasan conducts a demonstration for a visiting group of medical students, showcasing the use of microbots in GI procedures. She explains that these are tiny, ingestible robots that navigate the GI tract using pump jets, capturing high-resolution images and performing minor diagnostic tasks. The microbot provides real-time live video to the doctor, controlled via a touchscreen, enabling quick and accessible stomach examinations.

Dr. Grasan highlights the current applications of microbots, particularly in diagnosing conditions like gastritis, and their potential to replace some traditional endoscopies. She discusses future advancements, including lab-on-chip technology for taking samples and microsurgery capabilities for procedures like polypectomies.

During the demonstration, Dr. Grasan navigates the microbot through a simulated GI tract, showing the students intuitive controls and emphasizing the importance of data analytics. She also showcases a robotic-assisted procedure, where the robot performs intricate movements with precision, assisting in minimally invasive surgery. This hands-on experience gives the students a glimpse into the future of gastroenterology, inspiring them with the potential of cutting-edge innovations in healthcare.

WORK-LIFE BALANCE

Throughout the day, Dr. Grasan uses various health apps to streamline her workflow. These include a patient management app for scheduling and follow-ups, a telemedicine app for virtual consultations, and a research app that integrates with clinical trial databases, providing real-time updates on ongoing studies and new findings.

By 5:45 p.m., Dr. Grasan wraps up her clinical responsibilities and heads home. As she starts her drive, she receives an alert that Mr. Reynolds is reporting a significant reduction in pain. She smiles, reflecting on the power of integrative and personalized care.

EVENING AND PERSONAL TIME

Once home, Dr. Grasan enjoys dinner with her family, engaging in meaningful conversations and unwinding from the day. Her integration of advanced technologies—AI, robotics, microbiome analysis, non-invasive molecular diagnostics, microbots, the stool app, electronic nose devices, and digital avatars—not only enhances patient outcomes and contributes to clinical research but also allows her to maintain a healthy work-life balance.

As she spends quality time with her loved ones, she feels confident in the future of gastroenterology and the role she plays in shaping it.

(Many of the advancements and technologies described in the above narrative are either currently available or actively being developed.)

CHEERS TO THE SUMMITEERS

One day, my dad shared with me a parable he'd read in the newspaper about a frog that made it to the top of a summit despite many naysayers. It goes like this.

Once upon a time, a group of frogs decided to hold a competition to see who could climb to the top of a tall mountain. The challenge was tough, and the mountain was steep and treacherous.

The frogs began their ascent, and a large crowd gathered to watch. As the climb progressed, the spectators began to murmur and shout. "It's too difficult!" "They'll never make it!" "Why even try?"

One by one, the frogs began to lose hope. The negative comments from the crowd filled their minds, and they started to drop out of the race, turning back down the mountain. Soon, all the frogs had given up—except one.

This lone frog continued to climb, seemingly unaffected by the jeers and doubts of the onlookers. Step by step, it made its way up the mountain, never faltering. Eventually, against all odds, the frog reached the summit.

When the victorious frog returned to the base, the other frogs were amazed and asked, "How did you do it? How did you ignore all the negative comments and keep going?"

It was then that they realized the secret: the frog was wearing headphones with blaring music and hadn't heard a single word of the negative feedback. It simply focused on the climb and believed in its ability to reach the top.

This parable resonates with me because it spells out a seriousness of purpose and the power of disregarding detractors. It transports me back to when I rode solo through the Himalayas, listening to the silent voice of my heart. It reminds me of the many naysayers I encountered before, during, and after the trip—people who, despite never having undertaken such a journey, felt qualified to give advice. These naysayers included family members, friends, and even strangers. Expectedly, that advice extended to everything from how to run my business to how I should live my life. After a point, I stopped explaining myself and simply did what I felt was right for me. Because it wasn't about them; it was about me and my endeavor to find my path—to discovering my true self hidden beneath layers of societal norms. Celebrating the frog that shut out the noise, listened to its own tune, and made it to the top of the mountain is a good reminder that you, I, and anyone else *can* carve out a new future on our own terms if we truly want to.

THE SHIFT WITHIN

Perhaps it was easy for the frog to find focus amidst chaos, but how do *we* go about it? Carving out a new future on our terms is not just about survival or outward expressions of success. It's also about dealing with the loss of meaning and purpose in our work and beyond. In Chapter 2, we talked extensively about the massive transformative purpose (MTP) as a driver for making an impact in the world. But there's also an *inner purpose* that spiritual teachers like Eckhart Tolle talk about. Something that can awaken us from our disillusioned state and guide us *inside-out*—provided we are willing to listen.

The loss of meaning we face in our everyday lives is due to a lack of alignment between what we aspire for *outwardly* and what goes on *inside*.

For most of us in healthcare, our work was originally inspired by a deeper calling. That inner drive is now relegated more and more to the background, especially when our days are engulfed with *busyness*. Perhaps the calling was to heal, to care, or to provide comfort. But our days are now packed with business decisions to make, technology to learn, administrators to handle, business partners to convince, colleagues to motivate, regulations to watch out for, mortgages to pay, kids' college education to save for, better cars and houses to buy, and personal relationships to nurture. Under such external stresses of living, we've forgotten why we started this journey in the first place. We fill our time and space with running after *more*—more titles, more acquisitions, more rewards, and so on. These fillers may offer temporary respite and mask the feeling of meaninglessness. Maybe they offer some validation that it's all been worth it, but they won't make it really go away. The feeling will return. As we explored in Chapter 3, AI is at the cusp of magnifying everything, including this ongoing feeling of disillusionment.

However, there is a way to escape this quagmire. That way is inward. Unlearn the need for external validation and embrace the stillness within. Slowly, in the silence, *sense* an ever-present voice reverberating inside. That inner voice holds the key to our inner purpose. Ultimately, it reminds us of who we really are. When that inner essence flows into our outer purpose—whatever we pursue will be powered with higher clarity and decisiveness. It will make us unwilling to tolerate the current state of the healthcare system and help us expand beyond the limits we set for ourselves. By transcending the grip external forces have on us, we will experience a deeper sense of freedom. We will simply take empowered action—not tomorrow, but today—because our future is embedded in the quality of the actions we take now. That's the shift that must happen. The shift within.

ONE FINAL QUESTION

By now, it should be clear that the shift is both necessary and imminent. But perhaps you're wondering where or how to begin. One way to chart your course is by understanding what others have done. The upcoming sections are dedicated to that—stories of changemakers. You'll find it fascinating to uncover answers to the questions that still might be circling in your mind directly from those who have taken the leap. They not only understand their current position but also recognize its evolving nature. Whatever journey you decide to embark on after reading their stories is, of course, entirely up to you. Our world will always make space for the future to coexist with the past. New industries will form from the ashes of the old. Disruptors will operate alongside those who are disrupted. What matters is deciding which side you'd like to be on.

Reflect on this final question: *Are you just going to watch the digital transformation happen from the sidelines? Or will you be part of the force that shapes the future?*

SECTION: VOICES I - DIAGNOSTICS AND DEVICES

This section features interviews with innovators in diagnostics and medical devices, showcasing how new technologies are revolutionizing disease detection and patient care.

DIAGNOSING DISEASE THROUGH BREATH: AONGHUS SHORTT OF FOODMARBLE

Aonghus Shortt is the founder and CEO of FoodMarble. He has a background in engineering and data science. FoodMarble created the world's first personal digestive breath tester.

Themes: *personal motivation, IBS, breath analysis, hydrogen measurement, remote patient monitoring, digital device innovation*

Date: *May 2022*

Three Takeaways:

1. **Innovation Driven by Personal Motivation:**
   - Aonghus's journey into gastroenterology was inspired by his personal experience with his wife's struggle with IBS. His motivation to help her led to the development of a breath analysis device, showcasing how personal connections can drive significant innovations in healthcare.

2. **Advancements in Digital Health Technologies:**
   - The development and success of the breath analysis device illustrate the shift toward digital health solutions. By focusing on measuring hydrogen, methane, and hydrogen sulfide levels, Aonghus's work aligns with the broader trend of leveraging digital tools to improve diagnostic accuracy and patient care in gastroenterology.

3. **Growth and Future Vision in Digital Gastroenterology:**
   - Aonghus's company has seen substantial growth, reflecting the potential and demand for digital health innovations. His vision for the future includes expanding the use of their technology for continuous, real-time patient monitoring, which aligns with the book's emphasis on the exponential growth and impact of digital health technologies.

Praveen: Aonghus, you have a PhD in engineering in a completely different field, and you're a data scientist. How did someone with your background get into gastroenterology?

Aonghus Shortt: It's a good question. It was actually my wife, who was my girlfriend at the time. She has IBS and was struggling a lot. She'd seen various clinicians, including primary care doctors and gastroenterologists, and tried numerous medications. Despite undergoing several procedures, nothing significant was found. It's a common story where people end up with an IBS diagnosis.

I started doing some research since I had access to the literature. At that time, breath analysis had been used for a while with large benchtop devices. I noticed that the low FODMAP diet was emerging in the literature. This approach involves identifying foods that ferment rapidly into gases in the gut and reducing their consumption to feel better.

I found it remarkable because early trials showed significant improvement in as many as three or four people, who tended to feel better. Breath analysis was used in that research, which made me think: could I build one of these devices for Grace? When she ate certain foods, her breath hydrogen levels would rise significantly, indicating those foods might need to be limited in her diet. This allowed her to personalize her diet, which was pretty cool. That was the inspiration for what we're doing today.

Praveen: From my understanding, why did you choose to focus on hydrogen as a gas? Can you explain why hydrogen, in particular?

Aonghus: Yeah, there are a few gases that are relevant to the breath, with hydrogen being the primary one. When you eat something that isn't absorbed or fully digested, it reaches the microbes in the gut—usually in the large intestine—where it starts to ferment. This fermentation process by bacteria and other microbes produces hydrogen, carbon dioxide, and various other metabolites.

Hydrogen is the main gas produced, though some of it can be converted into methane or hydrogen sulfide. For our first-generation device, we chose to measure hydrogen because it is the primary gas of interest and is highly responsive to the food people eat. If you're not digesting food effectively, you'll often see significant increases in hydrogen levels. Conversely, if hydrogen levels don't rise, it suggests the food might be okay for that person.

For our second-generation device, we are also measuring methane and plan to release an update that includes hydrogen sulfide measurement. Our sensing array in the second-generation device can measure all three gases.

Praveen: What is in the device that listens to the hydrogen signal? How exactly does it sense?

Aonghus: Inside the device, there's a sensing canal. When a person exhales into the device, their breath passes through this canal, where a sensing array is located. This array contains multiple sensors that measure the electrical resistance of specific sensors. When a hydrogen molecule comes into contact with a sensor, it temporarily attaches to and then detaches from the sensor's surface, causing a reaction.

These reactions generate signals, and we collect multiple signals from the sensing array. We then use various models to translate these signals into concentration levels, which we can display to both clinicians and patients.

Praveen: Sounds fascinating. Now, can you share some numbers? How far have you come?

Aonghus: Yeah, we've sold over 30,000 devices so far, most of them directly to consumers. We started offering the device directly to consumers at the end of 2018 through our website. More recently, in the middle of last year, we launched our first medical device, called MedAIRE, which is an FDA class one device now available in the US.

In terms of other numbers, we're a team of 25 based in Dublin, though we're often in the US. We've raised over $6 million in VC funding. We've seen steady growth, with our sales doubling last year compared to the previous year, and we're seeing strong growth again this year. Both consumers and clinicians are very interested.

Often, consumers buy the device, gather data, and then bring it to their gastroenterologist. To support this, about a year ago, we decided to build a system that provides a dashboard for clinicians to see the results. By having a medical device, clinicians can use it in their practice, making it a source of revenue as well.

Praveen: Can you talk a little bit more about that? How could a clinician earn money by partnering with you? And how could it be a source of revenue for them?

Aonghus: Yeah, basically, if a clinician gets set up with us, they can order devices that are sent directly to their patients, or they can receive a bulk number of units to have on hand. This allows the clinician to give the device directly to the patient during office visits. Additionally, clinicians can avail reimbursement for the breath tests using the same reimbursement codes as traditional breath tests.

We're also increasingly focusing on remote physiologic monitoring (RPM). Conditions like IBS, SIBO, and functional constipation benefit from tracking the patient over time. For instance, in cases of functional constipation, measuring methane levels is very relevant as it tends to correlate with constipation in many patients.

We offer different models depending on the clinician's needs, but fundamentally, we provide a dashboard for clinicians to review patient results. This can guide patients through various breath tests, such as SIBO breath tests, tests for different food intolerances, or remote monitoring. We aim to make it easy for clinicians and their staff to set up and interact with patients, facilitating breath testing from home.

Praveen: Does this come under the same category as other remote patient monitoring devices, such as a blood pressure cuff or devices to measure diabetes that send data to the clinician? Would it come under the same category?

Aonghus: Yeah, exactly. From a reimbursement perspective, it's the exact same code.

Praveen: Yes, from a reimbursement standpoint.

Aonghus: Yeah, it's a very similar concept as well. You're tracking data that's relevant to the course of treatment. For example, a GI might initially diagnose a patient with SIBO using a conventional fasting morning breath test. This test can be done remotely with our device. If SIBO is detected, the patient is often treated with rifaximin, an antibiotic. During and after treatment, you can monitor the patient's hydrogen and methane levels.

This monitoring is crucial because you want to see if the treatment worked and if another course is required. Additionally, in about half of the cases, SIBO tends to return after treatment. Clinicians can monitor patients to see if symptoms return, if another course of treatment is needed, or to determine the best next approach.

For many conditions like SIBO, different treatment strategies might be needed. Identifying and resolving the underlying cause is crucial. For instance, if a patient has very slow motility, using a prokinetic agent might be beneficial. Tracking fermentation levels in real-time over a period can significantly aid in effective treatment.

Praveen: Now you have the benefit of more than one clinical study, isn't that right?

Aonghus: Yeah, we've done a number of studies validating the device itself and, more recently, exploring interesting ways to use the data.

Praveen: So, coming back to the business aspect, are you in touch with insurance companies? What are they saying in the US?

Aonghus: Yeah, from a business point of view, this is something we want to develop further because we haven't had discussions with payers yet. IBS is the number one diagnosis in gastroenterology. Looking at different cohorts of patients, like IBS-C (constipation-predominant), I saw numbers just before the call showing an average extra cost of $4,000 per patient per year. While this isn't as high as some other conditions, the large number of affected people makes it a significant cost for payers.

Considering the various expensive drugs used, we believe we can offer a tool that saves a lot of money. With the shift toward value-based care, our approach aligns well with these new healthcare models.

Praveen: So you've raised $6 million and sold 30,000 devices, and two-thirds of that is in the US. What happens next, short-term and long-term? What are your growth plans?

Aonghus: Yeah, there are some really interesting developments with our research collaborators. We're currently conducting a clinical trial at Johns Hopkins, for example. This trial looks at using breath readings over a period of time from the home rather than a single snapshot, such as a fasting morning breath test. Patients record their meals and take breath tests throughout the day, especially after eating. This approach is convenient because it allows patients to measure in a more natural setting.

We found that this method is particularly effective for predicting responses to rifaximin, a drug approved for IBS-D. This continuous data collection seems more effective than conventional testing, mainly because the digestive tract has a lot of variability, and many factors affect digestion. By capturing more data, we can improve diagnosis and treatment guidance.

Praveen: Do you have a vision for the future of GI? If everything happens like you think it should, what does that look like?

Aonghus: Ideally, for the many patients coming into gastroenterology—potentially half of them—who have conditions relevant to what we're doing, we could provide technology that allows them to gather data over time at home and use that data to manage their treatment. For instance, patients could identify which over-the-counter supplements help them or determine which medications would be most beneficial with their clinician's guidance.

This approach could also be purely food-related, such as deciding if a patient is a good candidate for a low FODMAP diet or another suitable diet. Guiding patients remotely through this process over time would be cost-effective and much more effective for both patients and clinicians. It would also bring satisfaction to clinicians by enabling them to treat these complex conditions more effectively.

Praveen: Aonghus Shortt, thank you so much for joining us today on The Scope Forward Show. I'm excited about what you're building. I always admire and respect innovators. What you're doing is fantastic. I wish you and your team great success. Thank you once again.

Aonghus: Thanks, Praveen.

Update from Aonghus Shortt (July 2024):

Since the interview, we've started to scale in US gastroenterology. We've got some major health systems and GI groups as customers now. We're also demonstrating the benefit of ongoing monitoring through remote patient monitoring (RPM). For RPM, we're particularly focused on recurring SIBO and chronic constipation and helping patients with dietary changes. We're seeing where this ties into less obvious conditions like Parkinson's, where we see that the biomarkers we measure could help guide treatment and ultimately help make this chronic condition less debilitating for patients.

TRACKING GUT'S ELECTRICAL ACTIVITY: DR. GREG O'GRADY OF ALIMETRY

Dr. Greg O'Grady is a general and gastrointestinal surgeon and Professor of Surgery at the University of Auckland, as well as the founding CEO of Alimetry. Alimetry is a MedTech company developing non-invasive diagnostic tools for gastrointestinal disorders.

Themes: *wearable medical devices, gut electrical activity, non-invasive diagnostics, data-driven healthcare, innovation in GI*

Date: *August 2022*

Three Takeaways:

1. **Innovative Diagnostic Tool:** Alimetry has developed a wearable, non-invasive device that measures the electrical activity of the gut, providing a new way to diagnose and understand gastrointestinal disorders. This innovation has the potential to fill a significant diagnostic gap in gastroenterology.
2. **Clinical and Commercial Viability:** The technology is supported by substantial research and development, including ten years of pre-commercial research and multiple scientific publications. The business model includes both a reusable kit and a consumable patch, making it feasible for integration into clinical practice and potentially profitable.
3. **Future of Gastroenterology:** The rapid advancements in technology, particularly in AI and big data, are expected to bring significant changes to gastroenterology. Alimetry aims to be at the forefront of this change, leveraging its

data-driven approach to improve diagnosis and treatment outcomes in the field.

Praveen: I'm curious to know more about your background and what led to the founding of your company.

Dr. Greg O'Grady: Alimetry builds wearable medical devices to diagnose gut disorders. We focus on disorders of function, like the software of the gut, how it behaves, works, and moves. My clinical background is in surgery, and I've been particularly interested in patients with difficult gastrointestinal symptoms that are hard to diagnose due to the lack of good tests. This interest, combined with my research on the electrical function of the gut and a passion for technology, led to the creation of Alimetry.

Praveen: So, can you talk a little bit more about this research? What kind of research and why electrical signals of the gut? Has this been known for a long time, or is this something relatively new in the industry or scientific community?

Dr. O'Grady: It has been known for a long time, but it's not well-known. It's been almost exactly 100 years since the electrical activity of the gut was discovered. The guy who discovered it, Alvarez, made a prophecy that one day, gastroenterologists would use electrical tools, just like cardiologists, to diagnose gut function. However, despite a lot of effort, it never happened. The stomach has a pacemaker and an electrical system that drives muscle contractions, but those signals are 100 times weaker than in the heart. Measuring them from the body surface is quite difficult. Much of my research, along with my team and colleagues, has been about tackling the technical problem of how to measure these weak signals accurately from the body surface to provide a useful clinical tool.

Praveen: Let's get to the basics a little bit. What is the stomach using the electrical signals for? We know what the heart does with them, but what does the stomach or the gut use them for?

Dr. O'Grady: It's relatively similar. The electrical waves drive the contractions. The muscle cells need that electrical signal to stimulate them to contract. It should be nice and regular, with rhythmic digestive waves that occur in your stomach every 20 seconds or so after you eat. But just like in the heart, you can get arrhythmias where it becomes irregular or fibrillation-type activities. In the stomach, you can get similar dysrhythmias that become very disorganized. These are the types of signals we're aiming to measure. You can also get ectopic pacemakers, where the waves of the stomach start traveling in the wrong direction. These signals correlate with diseases and symptoms.

Praveen: What kind of diseases?

Dr. O'Grady: Really common ones. About one in ten people experience gut symptoms after eating, and maybe half the time, it might be coming from the stomach. Conditions like chronic indigestion, functional dyspepsia, gastroparesis (where the stomach doesn't pump properly), and nausea and vomiting are the main things we're interested in.

Praveen: How big is this problem, globally or in the US?

Dr. O'Grady: It's really common, about one in ten globally, and a little higher in some places. For reasons we don't fully understand, it has been increasing at about three and a half percent per year over the last 20 years. Distress after eating is becoming more common, leading to significant suffering and health costs.

Praveen: What happens today when a patient suffers from digestive disorders such as dyspepsia? How is it diagnosed, or how does one know? Is it just based on patient complaints?

Dr. O'Grady: It's one of the most challenging areas in internal medicine because we lack a really good way to describe and diagnose these diseases. Patients often go through months or even years of a diagnostic odyssey, bouncing around and undergoing multiple tests that are always negative or inconclusive. It can be a really long journey, sometimes taking five years or more to reach a diagnosis. During this time, patients endure a lot of suffering, and clinicians struggle to manage these disorders. While indigestion symptoms might not be as severe, there's still a lot of negative testing and battles for clinicians in these cases.

Praveen: Let's talk about Alimetry. What does your device and solution do?

Dr. O'Grady: We measure the electrical activity of the gut from the body surface, which is completely non-invasive. This is a significant advantage as many gut tests involve tubes or radiation and can be unpleasant. The patient comes in fasted and sits in a chair, and we prep their skin before placing a wearable device on them. The innovation lies in our high-resolution device with 66 electrodes, as these weak signals from the gut are challenging to detect accurately. The wearable, wireless, sticky patch is placed on their abdomen, and they then eat a meal while logging their symptoms into our app. This helps us correlate the changes in the gut with their symptoms during the test. The data goes to the cloud, and we send a report back to the clinicians to interpret and guide patient care.

Praveen: Let's talk more about the electrical signals. I saw what the device looks like, and it's quite fascinating. You mentioned it captures these weak signals and then amplifies them. Could you categorize these signals?

Dr. O'Grady: We compile all these signals and form them into visual tools for clinicians. Unlike an ECG, where individual waveforms are analyzed, we process these signals into maps and visualizations. These can show whether the stomach's rhythm is regular or if it's scattered and irregular, indicating a neuromuscular problem where the nerves and cells driving contractions are failing. We can also detect when these rhythms or waves become spatially irregular, traveling in the wrong direction, which can lead to symptoms like bloating and pain after eating.

Praveen: And these sensors are housed in the device attached or spread throughout the white patch you apply over the abdomen?

Dr. O'Grady: Yes, it's quite cool technology. It's a printed stretchable circuit, similar to how you might screen print a T-shirt. We screen-print electrodes all over the patch, creating a high-density network. We then add hydrogel pads, like those used in ECG dots, to extract the weak signals, along with an adhesive. You peel it off and stick it on, with all these individual electrodes forming the patch.

Praveen: Fascinating. Let's talk about the business. Did you start in 2019?

Dr. O'Grady: Yes, we started in 2019, but we had about ten years of serious research before that. It wasn't until 2019 that we felt mature enough to bring this out as a product and company. We had around 100 scientific articles published before we found the right techniques. It was really hard work.

Praveen: Interesting. Can you share a bit about the funding situation? Did you raise money, and why did investors fund it?

Dr. O'Grady: We're a university spin-out company from the University of Auckland, which has a great tech transfer process. We spun out not just the IP and technology, patents, and algorithms but also a team of capable engineers. The university introduced us to investors interested in deep technology projects with rich IP. We met IP Group from the UK office, and in our most recent round, Movac led the investment. We also had support from the university and a few others, forming a syndicate to back this exciting New Zealand technology transitioning from academia to the real world.

Praveen: What is the business model for Alimetry in the US?

Dr. O'Grady: It's a pretty traditional business model for a diagnostic tool, similar to a pill cam or other GI function tests. Hospitals buy a kit that includes the reader device, and then there are consumables, which are the patches. These patches are single-use, so the hospital needs to purchase them for each test.

Praveen: Who would pay for it?

Dr. O'Grady: It's a traditional market setup where patients usually cover the costs. However, a significant challenge for medical device startups is not just regulatory approval but also securing reimbursement. Fortunately, there's a predicate technology that already has a reimbursement code in the US. Although that previous device wasn't particularly successful, our goal is to demonstrate the effectiveness and utility of our product to drive adoption. Once clinicians see the benefits and patients experience successful outcomes, we can focus on securing reimbursement to make it easier for practices to get paid for using the test.

Praveen: So private GI practices could potentially buy the device and bill for it?

Dr. O'Grady: Yes, absolutely.

Praveen: Do you see this as a platform of some sort? What's your view on the direction of where you're going with the business?

Dr. O'Grady: Yes, I do see it as a platform. We're currently focused on the stomach because it's an interesting organ that causes many problems and is relatively easier to measure. However, we're also interested in the colon, which can be measured from the body surface, though it's more complex. Additionally, the entire field of gut health is ripe for innovation. There hasn't been much advancement in this area, but with the dawn of new sensors, wearable technologies, data, and AI, there's a tremendous opportunity for innovative products to make a significant impact.

Praveen: Are you familiar with companies similar to yours, perhaps not in gastroenterology but other specialties that are trying to capture electrical signals and do something with them? Let's forget cardiology for a moment. But other than that, and of course, the brain.

Dr. O'Grady: Well, the heart and the brain are the two organs most obviously associated with electrical signals, but there are other smooth muscles in the body as well. We look at wearable patches and see companies doing this successfully with mature technologies. The brain is particularly interesting with many companies, even commercial ones, measuring brainwaves in various innovative ways. Some of these companies are super cool. We love seeing what people are doing with all types of wearables. As data geeks, we're excited by different sensors that can tell what your body is doing. We see ourselves as part of that wearable, data-driven community, even though we are a medical device company.

Praveen: If that is so, then why did you focus on getting reimbursed and working with clinicians? Why not make it a consumer device?

Dr. O'Grady: Firstly, there's a compelling need in the medical space. As a clinician, I find it crucial to fill a diagnostic gap that's been problematic for a long time and provide new approaches. The consumer market is more about the use case. While it might be interesting to measure your gut, will it improve your life? There's a big question mark. Many companies are looking at diets and how different foods affect the gut, which people are very interested in. However, can we provide that solution ourselves? I'm not sure yet. It's an interesting question, but for now, we're 100 percent focused on the medical space because it makes total sense. The consumer market is intriguing, though, and could be something for us or others to explore in the future.

Praveen: So if you start viewing yourself into the future, let's forward one year from now, three years from now, five years from now. How do you see the progression for Alimetry?

Dr. O'Grady: These new technologies go through several stages. As a clinician-led company, having seen many patients myself, our goal is to change the standard of care and genuinely address the clinical community's needs. I feel this need strongly as both a customer and a provider. We aim to make meaningful changes become the standard of care and provide answers that are currently lacking for many patients. If we achieve that, everything else will fall into place commercially. So far, so good. We have some exciting data coming out soon that I think will garner a lot of attention.

Praveen: Talking about the data that you've captured so far, what have you learned from all this data and the analysis? Are you doing any broader analysis from the accumulated data, and what have you learned from it?

Dr. O'Grady: One reason this field is exciting is that there's so much to learn. Unlike the heart, which is well-characterized, we can learn a lot even from studying normal, healthy people. We've generated data from many hundreds of patients, allowing us to formulate reference ranges and understand what normal digestive patterns look like after eating, the normal amplitude of contractions, and more. By defining what is normal, we can identify specific patterns and phenotypes in diseases that fall outside this range. With our app, we collect symptoms simultaneously, enabling us to make deep correlations with big data sets and determine which symptoms are associated with which patterns in different patients. This will only improve as we gather more data, helping us to make the best use of this tool.

Praveen: Greg, what's your take on the future of gastroenterology, or perhaps medicine as a whole?

Dr. O'Grady: It's an exciting time. It might sound like a cliché, but I believe we're about to undergo some major changes. Technology advances exponentially, and I think the rate of change will take people by surprise when AI and big data truly take off. Right now, there's some skepticism since these concepts have been discussed for so long, but the progress being made is significant and somewhat under the radar. I believe we're on the brink of a breakthrough, where the exponential rate of change will surprise many.

Gastroenterology won't be spared from this transformation. We're already seeing AI in endoscopy, and it won't be long until the power of data spills into other areas of gastroenterology. It's a super exciting time, and I love being on the side that promotes and brings about this change. That's certainly where we want to head as a company as well.

Praveen: Professor Greg O'Grady, it was fantastic to have this conversation with you. Any final comments before you take off?

Dr. O'Grady: Thanks for having me on your show. It's been fun to meet you and talk about these things. For those who are interested, feel free to get in touch. We'd love to work with more gastroenterologists on what we're doing.

BUILDING SMART TOILETS: DR. SONIA GREGO OF COPRATA

Dr. Sonia Grego is the cofounder and CEO of Coprata. Her background is in biomedical engineering, medical device development, and translational research. Coprata is pioneering smart toilet technology for non-invasive gut health monitoring and gastrointestinal disorder diagnostics.

Themes: *smart toilets, stool data analysis, digital health innovation, remote patient monitoring, IBS, IBD, constipation, digital devices, sensors, stool sampling*

Date: *June 2022*

Three Takeaways:

1. **Data Collection from a toilet:** Dr. Sonia Grego and her team have developed a smart sampling toilet that captures and analyzes stool data post-flush. This innovative approach addresses a critical gap in gastrointestinal health monitoring, providing valuable data that is typically challenging to collect.
2. **Potential for Broad Applications:** The smart toilet is designed to benefit patients with various GI conditions, such as IBS, functional constipation, and IBD, as well as health-conscious individuals. It also holds promise for use in research studies and clinical trials, offering objective data on bowel movements and frequent stool sampling that can enhance the accuracy and data richness of these studies.
3. **Go-to-market strategy:** The initial steps for the smart toilet include studies with users to demonstrate the healthcare value of the data generated by toilet-based gut

monitoring. Future expansions include stool sampling and medical-grade devices targeting specific populations and clinical research applications.

Praveen: Sonia, you've led the development of a smart sampling toilet that automatically captures stool data post-flush at Duke University. I'm sure that while growing up or studying engineering, you didn't dream of working on smart toilets. So how did this come about?

Dr. Sonia Grego: It's been an interesting journey. My PhD is in physics, and I have over 20 years of experience working in applied technology and engineering, developing biomedical technologies. About eight years ago, my colleagues and I at Duke University began to explore toilet technology as part of a larger center funded by the Gates Foundation.

If you think about it, the toilet is a fantastic appliance—very effective at removing waste, and it hasn't changed much since it was first introduced into homes at the beginning of the century. It's a white ceramic bowl with water, and you flush, and your waste goes away. While working on technologies for waste treatment, we asked ourselves if there was valuable data in this waste that we were flushing away. Could we capture it before it's flushed? The answer was a resounding yes.

That's what we set out to do—develop technology that analyzes stool data. In my experience with sanitation technologies for other environmental applications, we've deployed and tested many toilets with real users. We became aware of how sensitive the topic of using the toilet can be. It's a private, personal act, and users, particularly women, are very sensitive about the use of a toilet. So, when designing the Coprata smart toilet, we made a great engineering effort to ensure the product doesn't appear different to the user, thus avoiding discomfort.

Praveen: How did the exact idea come about? Were you simply thinking, "Hey, we got to measure stool data. There is a lot of data in stool, so let's figure out a device that helps us do that." What was the process of innovation?

Dr. Grego: Well, the process was rooted in our conversations with physicians, collaborators, and gastroenterologists here at Duke. We learned that there is indeed a lot of valuable data in stool, but it's very difficult to collect. Gastroenterologist collaborators told us that they spend most of their patient visits just trying to understand the regularity and characteristics of the patient's bowel movements. They said, "90 percent of the visit is just figuring out the consistency. What do they mean by the volume? Is it little? Is it a lot? What's little? What's a lot?" People don't have a frame of reference because it's a private act, and everybody only knows their own experience.

The physicians expressed frustration over the mismatch between what patients report and what clinicians understand. For instance, when patients say they go to the bathroom 20 times a day, is that really true? Is there something coming out 20 times? It seems unlikely. Additionally, when a stool sample is needed, patients are often sent home with a stool kit, but these are returned only about 40 percent of the time because people just don't want to do it.

The need for better stool data collection was clearly communicated to us. We realized that while it might be possible to engineer something that scoops down in the bowl or takes a picture of the stool, our experience told us that it wouldn't be practical. Users want the toilet to look and function normally. They just want to see a white bowl with water in it, no gizmos around.

Our laboratory at Duke is fully equipped to test toilets. We understand the physics, fluid dynamics, and engineering involved. We set out to figure out how to conduct stool analysis after the stool has left the bowl and outside the user's view. This was a complex challenge that took months of brainstorming, testing, and trial and error.

Praveen: How does this data get transmitted? I'm assuming it gets transmitted either via Bluetooth or the cloud. Can you explain that process?

Dr. Grego: Yes, the data is now being collected through imaging and analog sensors. Our first pilot is ongoing in our own facilities, so we're actively collecting and analyzing the data. The data is processed on our servers and coupled with algorithms to transform these signals into meaningful information parameters. Users will receive scores on various metrics like sitting time and changes from their baseline.

Praveen: So, is it Bluetooth or WiFi enabled? How is it being transmitted?

Dr. Grego: Currently, it is WiFi enabled for convenience, similar to how a smart TV connects to your home network. The data is then analyzed and relayed to an app.

Praveen: Just like setting up a smart TV, you configure the smart toilet to your home network, and all this data gets analyzed and relayed to an app.

Dr. Grego: Exactly. We envision the data being summarized in the form of a dashboard for user convenience. One common question we receive is about the longevity of these sensors. Once installed, the sensors will work indefinitely, capturing every bowel movement. With 100 toilet installations and one bowel movement per day, we could have 200,000 data points in a year. We believe Coprata will be a truly at-home biosensor for GI tracking that hasn't been developed before. The data will be summarized in a dashboard, both for users and clinicians interested in seeing it. For users, we envision an app and, for clinicians, some form of easily transferable information.

Praveen: Do you have the number of data points you've collected so far? What kind of data have you collected in terms of quantity?

Dr. Grego: I don't have the exact numbers off the top of my head, but we've developed our algorithm using around 3,000 images, which we obtained through crowdsourcing. We've published a study on our ability to sample stool from the toilet, which included hundreds of data points. We have a few months of data on the toilets we've installed. You can calculate multiple data points per user per day—up to five users a day, seven days a week, over a few months. So, we've collected a substantial amount of data to support our development and testing efforts.

Praveen: What have you learned so far?

Dr. Grego: We've learned that our algorithm is very good at recognizing stool consistency and is as reliable as a clinician. Studies in the literature show that algorithms can often be better than people at recognizing the characteristics of stool specimens. We are confident that sensor-based analysis of stool will provide clinicians with more accurate data than what a patient can observe on their own.

We have also learned that it is possible to sample stool from the toilet region and conduct biochemical analysis on that sample. While this capability is not yet included in our current product, it has been characterized separately with funding from the NIH. Our platform currently collects comprehensive information on bowel movements, but we plan to develop it further to include stool sampling. This will allow for fecal specimens to be sent to a laboratory for biochemical analysis in the future.

Praveen: What kind of disease conditions are you currently planning for?

Dr. Grego: The toilet would benefit patients with a variety of GI conditions that result in bowel irregularity, which is practically all of them. Particularly, patients with IBS, functional constipation, and types of IBD would immediately benefit from the toilet. We also believe that health-conscious individuals who are not currently GI patients would benefit. Anyone who uses an Oura ring to track their sleep or a WHOOP to monitor physical activity could benefit from tracking their bowel movements to see the effects of their diet and lifestyle. Regularity and gut health contribute significantly to overall health and happiness.

We also envision using our smart toilet in research studies and clinical trials. Clinical trials, especially for GI conditions, often rely on patient self-reporting for improvements in bowel movement frequency, straining, or urgency. Our technology would provide investigators and stakeholders with a robust set of objective data for these trials.

There is a lot of potential in toilet-based gut health monitoring. The reason this area has not been explored extensively is due to the technical difficulties. We chose to focus on stool analysis because of its complexity and variability. Urine analysis is simpler, as urine is a liquid and more homogenous. We consciously decided to tackle the more challenging problem of stool analysis, knowing that it would provide significant benefits. From an engineering standpoint, we believe we have solved this problem and are now focused on bringing a product to market that will benefit patients soon.

Praveen: What is the business model? Is it mainly by selling directly to consumers, or are there other types of revenue sources that you're envisioning?

Dr. Grego: Our initial step will be direct-to-consumer sales. We believe that as our user base expands, we'll be able to conduct studies demonstrating the value of the data our system produces for healthcare. Once we can show improvements in outcomes, time savings, and healthcare cost reductions, we envision shifting to a B2B2C model. In this model, payers and large employers would be interested in subsidizing or partially reimbursing the device for their patients, similar to what is done for diabetes management.

While diabetes is a chronic disease that has advanced considerably in terms of remote patient monitoring, GI chronic diseases are also expensive and significantly impact the quality of life for a large and growing number of people. We believe that large employers and payers will recognize the benefits of a system that enables remote patient monitoring, leading to further savings and improved patient care.

Looking further ahead, our market could expand into stool sampling and medical-grade devices targeting specific populations and comprehensive sample analysis. So, our model starts with a clear consumer focus but has the potential to expand into various healthcare and clinical research applications.

Praveen: One final question: what is the future of a specialty like gastroenterology in your view?

Dr. Grego: The future of gastroenterology, like many other specialties, lies in telemedicine. The push toward telemedicine and the development of digital technology are crucial. For gastroenterology, there has been no at-home biosensor tracking specific physiological activities of the patient until now. We believe our smart toilet will be an important tool that enables remote, proactive care. It will allow for long-term tracking to keep patients in remission and maintain their health.

Update from Dr. Sonia Grego (June 2024):

Since this interview in 2022, Coprata has advanced its product to include the capability of automated hands-free stool sampling. This groundbreaking achievement has shifted the business strategy to a focus on clinical applications of fecal biomarkers, including biomarker discovery and microbiome studies for GI diseases, oncology, and broader chronic disease applications.

BREAKING BARRIERS IN CANCER SCREENING THROUGH NON-INVASIVE DIAGNOSTICS: DR. ERICA BARNELL OF GENEOSCOPY

Dr. Erica Barnell is the cofounder and chief medical and science officer of Geneoscopy. Her background is in biological sciences, applied economics, and bioinformatics. Geneoscopy is a medical diagnostics company focused on developing non-invasive tests for gastrointestinal health, including colorectal cancer screening using stool-based RNA analysis.

Themes: *colorectal cancer screening, non-invasive diagnostics, RNA technology, liquid biopsy, AI in medicine*

Date: *December 2023*

Three Takeaways:

1. **Embracing Technological Advancements:** Dr. Erica Barnell emphasizes the importance of integrating innovation, technology, and AI into gastroenterology to improve patient care and reduce administrative burdens. This aligns with the book's focus on innovation driven by personal motivation and advancements in digital health technologies.
2. **Systemic Support for Physicians:** The burden of integrating new technologies and improving healthcare should not fall solely on physicians. It requires systemic changes and support from hospital administrators and CEOs to provide the necessary resources. This takeaway ties into the book's theme of growth and future vision in digital gastroenterology, highlighting the need for comprehensive support systems.

3. **Potential of Precision Medicine:** Dr. Erica's discussion on the advancements in precision medicine, particularly in IBD and oncology, underscores the transformative potential of personalized healthcare. This reflects the book's core message about the future vision of digital gastroenterology and the importance of precision medicine in improving patient outcomes.

Praveen: Geneoscopy has a new study published in JAMA (The Journal of the American Medical Association). What was the study about, and what was published?

Dr. Erica Barnell: Geneoscopy has been pretty laser-focused for the last couple of years on developing its lead assay, which is a non-invasive test to screen average-risk individuals over the age of 45 for both colorectal cancer and advanced adenoma. The last time we spoke, we discussed some of the preclinical trials we completed. Most recently, we underwent design lock of that assay, completed about 28 analytical validation studies, and finished our clinical validation study to support a pre-market approval application to the FDA. That application was submitted in January of this past year. The clinical validation study results were published in JAMA in October.

Praveen: What did the results say?

Dr. Barnell: It was really amazing. We recruited over 14,000 patients for this study. The data analyzed about 9,000 individuals who both completed a ColoSense test and subsequently had a colonoscopy. For individuals who had colorectal cancer detected on a colonoscopy, the ColoSense test demonstrated 94.4 percent sensitivity for colorectal cancer and 46 percent sensitivity for advanced adenomas, which are precancerous lesions most likely to undergo malignant transformation. These sensitivity profiles were achieved with an 88 percent specificity for no lesions on a colonoscopy. This means if you don't have any lesions, there's a 13 percent chance our test might come back positive, referring you to a colonoscopy.

Praveen: I want to ask a very objective question. There are a handful of non-invasive options available now, and how does the Geneoscopy test compare with the other options that exist today?

Dr. Barnell: Currently, three different modalities have been approved for colorectal cancer screening. The fecal immunochemical test (FIT) has been around for years, but it must be done annually, which affects patient compliance. Its sensitivity is only 74 percent for colorectal cancer and 24 percent for advanced adenomas, making it less effective compared to advanced molecular diagnostics. Another option is the Septin 9 test, a blood-based test with similar issues and lower sensitivity for colorectal neoplasia. The third test on the market is Cologuard, provided by Exact Sciences', which cites a 92 percent sensitivity for colorectal cancer and 42 percent for advanced precancerous lesions. However, a significant issue with Cologuard is the lack of data supporting sensitivity in the 45 to 49 age group. What we know is that its methylation-based biomarkers show reduced accuracy in younger patients. For example, while it cites a 42 percent sensitivity for advanced precancerous lesions in patients 50 and older, their sensitivity in the 45 to 49 age group is only 33 percent. As this younger age group has recently been added to the recommended screening population, we don't have sufficient data on how current tests perform for these subjects.

Praveen: If and when liquid biopsy becomes mainstream, what would happen to all these different tests?

Dr. Barnell: Our test is currently undergoing FDA approval as a pre-market approval application with a similar indication as those on the market. There's another test produced by Guardant, a liquid biopsy company, that was also submitted around the same time as ours. What we're seeing with current liquid biopsy technology is that the sensitivity for colorectal cancer and advanced adenomas is not high enough to justify the cost. For instance, a cost-effective analysis in JAMA recently indicated that for the accuracy profile cited from the ECLIPSE study, the test would need to cost about $50, whereas currently, they're charging around $1,000 per test. This discrepancy between accuracy and cost is a significant issue. Looking ahead, as technology advances, the accuracy of liquid biopsies will likely improve, especially for cancers where we lack effective screening methods like ovarian, pancreatic, and lung cancers. In these areas, liquid biopsy could be transformative. However, for cancers where we already have good screening modalities, such as colorectal, cervical, and breast cancer, the current accuracy profiles of liquid biopsies might not justify their high costs.

Praveen: Wouldn't that change in the future? There are also GRAIL, Freenome, and several other liquid biopsy companies in the pipeline.

Dr. Barnell: Absolutely. The GRAIL-Galleri test has been on the market for two years now, and they've done about 100,000 tests. There needs to be a change in policy for these tests to be widely adopted, as well as a better understanding of how to demonstrate clinical validity. The FDA is working on new guidelines to integrate these tests into clinical practice, but currently, we lack sufficient data on their impact on clinical outcomes. With improvements in accuracy and policy changes, these tests will likely become mainstream. Whether that happens in 5, 10, 15, or 20 years is hard to predict, but science is always advancing to help patients.

Praveen: Let's say that it does get mainstream. How would that impact what you do?

Dr. Barnell: If a liquid biopsy achieves comparable sensitivity to our test—50 to 60 percent for advanced adenomas and over 95 percent for colorectal cancer—the current version of our lead assay could become obsolete, which is fine because that's how evolution works. However, I don't see that happening in the near future. This is why Geneoscopy is committed to continuous innovation. While we've focused on colorectal cancer screening over the last five years, we've also been developing our pipeline in other areas of gastrointestinal health. I'm particularly excited about our work in IBD, which needs more precision medicine to help both physicians and patients. The field is currently confusing unless you're at a large academic center with many resources. We aim to improve our pipeline and find new ways to impact the field.

Praveen: Let's rewind the clock a little bit, Erica. How did you get started? What's the backstory behind Geneoscopy?

Dr. Barnell: I started my career in the gastrointestinal field as an undergraduate researcher at Cornell, but I was conducting research in my hometown, St. Louis. We were working on diagnosing environmental enteropathy in children in Africa—a non-specific gut inflammation that causes growth faltering. The condition is asymptomatic, so identifying affected children for intervention was challenging. Since we couldn't scope kids in Africa due to limited resources, we had moms send us diapers. We developed a technology to isolate human cells from the diapers, analyze the transcriptome, and identify which children had the disease.

I thought this technology was amazing and wondered why we weren't using it in the US for conditions like colon cancer, IBD, IBS, and celiac disease. When I suggested this to my PI, he supported the idea but didn't want to pursue it himself. So, when I joined the MD-PhD program, I decided to pull the technology out of the university and founded Geneoscopy. That's how it all started.

Praveen: Interesting. And you've started the company with your brother. How's that going?

Dr. Barnell: It's so funny, Praveen. When I tell people that I started a company with my brother, they either say, "I could never start a company with my sibling," or, "That sounds like so much fun." For me, it's the latter. I truly love working with my brother. He has an MBA from Wharton and has four years of experience in healthcare investment banking and private equity. Together, we make a pretty unstoppable force—him on the business side and me on the science side.

Entrepreneurship is volatile. You can start the day excited about something, and by the end of the day, you're dealing with multiple fires. Having someone I trust intrinsically, who keeps me accountable and who I enjoy working with, has been incredibly helpful in ensuring our success at Geneoscopy.

Praveen: Just so we know the kind of business impact that the company has made, can you give us an update on how much money Geneoscopy has raised?

Dr. Barnell: Totally. We've gone through the entire gauntlet of fundraising. We started with $20,000 of founder seed money, did a $1 million friends and family round, then about a $12 million Series A, and last year, we raised a $105 million Series B, which was a combination of debt and equity. We're currently raising our Series C, which will be about $100 million, to support the launch of our ColoSense test as well as our pipeline work in IBD. So, we've definitely followed what they recommend in business school for raising funds.

Praveen: Very exciting. Congratulations to all of you for pulling this off. Why are the investors funding companies such as yours? What's the underlying thesis here?

Dr. Barnell: A lot of people talk about investors backing good ideas, but I think it's more about execution. Geneoscopy is a great idea, and what we're doing is crucial and will save lives. But investors who back us see how well we execute. We're conscious of the environment and external factors, and we pivot effectively and leverage our resources to succeed.

Investors tend to invest in people they trust to get things done. Inevitably, things happen that can't be anticipated or controlled—like the COVID pandemic. You have to be able to pivot and succeed despite obstacles. For instance, during our clinical validation study in the pandemic, we faced unprecedented challenges. There was no guidebook for executing a clinical validation study during a pandemic, so we had to develop a novel way to recruit patients. We did it cost-effectively and completed recruitment in under twelve months. Necessity truly is the mother of invention.

Being creative, agile, and determined has been crucial in getting investors on board and believing in our mission.

Praveen: What's the next $100 million of funds being invested in?

Dr. Barnell: Series C funds will be dedicated to three main aspects. First, as we move toward FDA approval of the ColoSense test, the funds will support commercialization efforts. Second, we're building out our pipeline in the IBD space. We're developing a test to evaluate disease activity in patients with IBD and to predict therapeutic responses to certain biologics for these patients. The funds will support the product development cycle for this test. Third, we've attracted interest from pharmaceutical companies wanting to integrate our platform technology into their phase trials for therapeutic approval. We recently announced a partnership with Adiso Therapeutics and have several ongoing pilot studies with big pharma for clinical trial integration. These three efforts will be pursued in parallel over the next couple of years.

Praveen: Let's get to the science, Erica, and that's your home turf. If you were to explain this to a ten-year-old, what is the science behind your technology, starting with why you chose RNA?

Dr. Barnell: Absolutely. There are two components that make Geneoscopy technology so accurate. First, we focus on stool samples. These samples are a direct source of the cells that indicate disease. We're gathering exfoliated epithelial cells from colon cancer or from the inflammation that causes IBD, making it a bit like a noninvasive biopsy. Blood, on the other hand, is more dispersed, making it harder to find those cells—it's like looking for a needle in a haystack. With stool samples, we can more easily identify the disease signals.

Second, there's the choice between DNA, RNA, and proteins. Think of DNA as the cookbook that contains all the recipes. RNA is like the ingredients you gather after reading the cookbook, and the protein is the final cake you bake. RNA is very dynamic and reflects what's happening in the cell in real-time. It can change quickly in response to events like cancer, inflammation, or other cellular changes. This makes RNA a better marker for detecting these changes as they occur.

Specifically for our ColoSense test, RNA is better at detecting disease in younger populations compared to DNA, especially methylation-based biomarkers. Methylation is an age-related process, which means that methylation occurs more frequently in older individuals. RNA is likely to be more sensitive in detecting cancer and precancerous lesions in younger patients. Beyond oncology, RNA is crucial for detecting diseases like IBD, IBS, and celiac disease, where there are no DNA changes to indicate disease. RNA is the biomarker that can accurately reflect these conditions.

Praveen: This might come across as a naive question. How did you pick colon cancer? When I hear you describe this, could this not apply to other types of cancers?

Dr. Barnell: I think it was an opportunity that presented itself based on what I was seeing in the clinic. When I started the MD PhD program, my first rotation involved a woman who came in with stage four colorectal cancer. She was over 50, so I asked why she hadn't gotten a colonoscopy. She said she didn't have time to take off work and had kids at home. It was devastating to hear that. I thought about the technology I was developing for kids in Africa and realized it could help someone like her. That inspired our lead assay, ColoSense.

But I do believe this is a platform technology. I'm excited to return to my inflammatory disease research roots, where I started. As the ColoSense test moves toward commercialization, we'll be looking at IBD and IBS. We're also considering other GI cancers like pancreatic or gastric cancer. Beyond GI, this platform could be applied to cancers in the GU system, such as transitional cancer and potentially ovarian or endometrial cancer.

Praveen: Erica, let's talk about the future. What is your vision of the future of gastroenterology, the future of medicine? Where's all this going?

Dr. Barnell: The future is bright. Every field of medicine is currently gaining traction in integrating innovation, technology, and AI to help physicians do their jobs better and reduce their administrative burden, allowing them to focus on patient care. In GI, we're seeing this integration with AI-enhanced scopes and novel diagnostic technologies, whether RNA-based, DNA-based, or methylation-based.

Specifically for IBD, we're at a stage similar to where oncology was ten years ago. During my PhD, the focus was on integrating precision medicine into oncology. We saw the emergence of assays like FoundationOne, which revolutionized cancer treatment by targeting therapies based on the genetic profile of the cancer rather than its location. This approach has transformed how we treat cancer.

For IBD, we are now seeing a significant effort in developing tests and therapeutics. There's a lot of investment in this area, and we want to be part of that innovation. Our goal is to develop a test that can predict which therapy would be most effective for an individual patient. By integrating into pharmaceutical clinical trials, we can develop companion diagnostics for new therapies, much like what Foundation Medicine did for oncology.

In the next five years, I envision us creating these companion diagnostics, effectively matching patients with the most effective treatments based on their unique profiles. This approach will enhance precision medicine in GI and improve patient outcomes significantly.

Praveen: As we wrap up our conversation, here's a final question. Should a private GI physician be worried or excited about this future?

Dr. Barnell: Excited.

Praveen: I'm sure you'll say that. Let me frame the question differently. Who should be worried, and who should be excited?

Dr. Barnell: Sam Altman, the CEO of OpenAI, went to my high school, and he gave a talk there a couple of months ago. He said it best: everyone should be excited about the future because it makes our lives easier and our jobs simpler. Predicting where we're going to be in the future is impossible, and I don't know where AI will be in five or even two years from now, but I do know that it will help improve our lives significantly.

The burnout in healthcare is so great because of the massive administrative burden imposed on us to document everything. As we continue to integrate AI into documentation and electronic health records, it's going to make physicians' lives a lot easier and more tolerable, allowing them to get home to their families on time without feeling like they're sacrificing their patients' healthcare.

The people who need to be worried are those who can't embrace change, like the Blockbusters, Kinkos, and Kodaks of the world. If you're unable to see how these advancements can benefit you and refuse to integrate them, it's going to be a really difficult transition. Those who don't embrace AI and new technologies will lag and fall behind, which will be challenging for individuals and companies built on outdated systems.

Praveen: Someone out there could be thinking, "Hey, it's easy for you to say as a scientist and researcher doing this as a day job. I have to meet RVUs, be in the endoscopy room, and fill my EHR. How do I stay relevant?"

Dr. Barnell: The burden might not be on the physician alone. From a regulatory and systemic standpoint, and this is something medicine is not great at doing—we still have beepers and fax machines—there needs to be systemic change to ensure support for physicians. Hospital systems need to provide scopes with AI integrated into them so that physicians can better detect polyps and maintain a comparable ADR to other GIs who have embraced technology. The responsibility falls on hospital administrators and CEOs to make sure their physicians have the necessary resources to do their jobs effectively. While it's a big lift, my hope is that people will start to embrace these changes.

Praveen: Excellent. Erica, this was fantastic. Thank you.

Dr. Barnell: Thank you so much for having me. I love speaking with you, and your show is amazing. I really appreciate being on here.

DEMOCRATIZING ENDOSCOPIC ULTRASOUND TECHNOLOGY: DR. STEPHEN STEINBERG OF ENDOSOUND

Dr. Stephen Steinberg is the founder and chief medical officer of EndoSound. His 40-year career focused on the use of endoscopes to perform diagnostic and therapeutic gastrointestinal procedures. EndoSound is a medical device company that transforms any flexible upper endoscope into a fully functional endoscopic ultrasound (EUS) scope, expanding access to EUS technology globally.

Themes: *endoscopic ultrasound, affordable imaging, 3D printing, medical device innovation, medical breakthroughs, commercialization*

Date: *February 2024*

Three Takeaways:

1. **Democratizing Medical Technology**: By developing a cost-effective solution for endoscopic ultrasounds that can be used with standard endoscopes, EndoSound aims to democratize access to advanced imaging technology, particularly benefiting surgicenters, non-academic and smaller healthcare facilities and second and third world countries.
2. **Overcoming Technological and Financial Barriers**: From initial self-funding and seeking investment through friends and family to obtaining breakthrough designation from the FDA, Dr. Steinberg's experience underscores the importance of resilience, strategic partnerships, and navigating regulatory landscapes in advancing healthcare technologies.

3. **Expanding Clinical Capabilities and Professional Satisfaction**: By providing general gastroenterologists with advanced imaging tools, the innovation not only improves clinical outcomes but also offers greater professional satisfaction by enabling more comprehensive patient evaluations and interventions.

Praveen: What excites me is to watch the pace of innovation that's happening in gastroenterology and leaders such as you who've been on the clinical side and who are stepping up into innovation and creating this next wave of medical devices. Before understanding your innovation, what is endoscopic ultrasound?

Dr. Stephen Steinberg: Endoscopic ultrasound is an endoscopy technology. Everybody, I think, is familiar with endoscopy, the ability to use a flexible scope through somebody's mouth to look at the esophagus and the stomach and parts of their small intestine. Surely, people are familiar with surface ultrasound, Butterfly iQ+™, and other technologies that allow us to look at the fetus, organs, and areas of the body from the outside. An endoscopic ultrasound simply takes that same technology and looks from the inside out. And because we're located at the end of a scope and can access areas not reachable with surface ultrasound, we get a unique view and have the ability to perform interventions in otherwise inaccessible parts of the body. It's basically an imaging technology, like MRI or CT. And the real concept is that if you can see something, you can do something.

Praveen: How did you come about with this innovation? What's the backstory

Dr. Steinberg: About ten years ago, I realized that the emerging technology in endoscopy, EUS, was becoming more and more useful but was limited in availability due to its high cost, around half a million dollars. This made it inaccessible despite its growing utility. I wanted to democratize endoscopic ultrasound, making it available to those without a large capital budget.

After spending years trying to pitch my idea to big companies with little success, a serendipitous event in 2015 changed everything. My wife, the provost and COO at Oregon Health Sciences, invited me to a meeting with a startup called Sonivate Medical, which was developing an ultrasound probe for battlefield triage. They were putting an ultrasound probe on the tip of a finger. Seeing this, I wondered if we could put it on the end of a scope.

Scott Corbett, the ultrasound engineer from Sonivate, and I started discussing the idea. It took us about two years to develop a design that allowed us to incorporate ultrasound onto a flexible endoscope while enabling interventions. After another six years of development and FDA approval, we finally released our product on December 27 of last year. Meeting Scott was a turning point, and without it, my idea might still be on the drawing board.

Praveen: That's an amazing story. And also, congratulations on the FDA. I saw the announcement come out in Jan, so that's exciting. Is the product now commercial?

Dr. Steinberg: Yes, we are currently in the launch phase. We're doing a limited launch to understand how people will use it and the best ways to set it up in a surgical center or hospital. We have a similar limited launch in Latin America. The goal is to ensure we find the right niche, provide the most value, and position it for easy adoption by the general population of gastroenterologists.

Praveen: What were the biggest bottlenecks you faced to get it to the point of receiving the first breakthrough designation from the FDA in 2021?

Dr. Steinberg: The biggest bottleneck was the complexity of the technology. We weren't just extruding something out of plastic; we were dealing with electronic and ultrasound components that had to be pieced together. Fortunately, my partner Scott had connections from his career and his work with Sonivate, which helped. However, it was still challenging from both an engineering and financial perspective.

We also had some fortuitous timing. The issues with the duodenoscope and infection transmission arose around the same time. The FDA was very concerned about the transmission of highly pathogenic bacteria and realized that the same problem existed with the linear echo (EUS) endoscopes. They were very interested in our alternative that could eliminate the design problem and make the technology safer and more accessible. This interest led to our breakthrough designation. So, it was a combination of technical challenges and fortunate timing with the FDA's focus on infection risks.

Praveen: Steve, I often like to explore exponential technologies, which are advanced technologies multiplying as a force. For your innovation to succeed, it seems imaging had to advance independently, sensor technology had to advance, and mechanical parts becoming smaller had to happen. I want to explore with you what broader innovations in technology and medicine, even beyond medicine, enabled your innovation to come through.

Dr. Steinberg: The one that comes immediately to mind is advances in 3D printing. The challenge with dealing with plastics and disposables is how to iterate efficiently. Science is about getting it less wrong each time until you get closer to the truth. Iteration can be expensive and time-consuming, but 3D printing has evolved to allow for quick and relatively inexpensive iterations. You can make a change and have a new part by the afternoon instead of waiting weeks or months for new molds.

In our field of ultrasound, single-crystal technology has provided significant image enhancements, allowing for smaller devices with better imaging than legacy systems. Advances like those seen in Butterfly iQ+™ will bring ultrasound chips down to the pennies, potentially making a totally disposable alternative. However, 3D printing has been the most transformative, making it easier to implement ideas and make changes quickly.

Praveen: Did you have a 3D printer in your lab, or did you work somewhere where one was available to you?

Dr. Steinberg: In the early days, our first engineer, knowledgeable in 3D printing, was sending work out to a nearby lab with a 3D printer. After a few weeks of hearing him say, "I've got this idea; I've got to make a change," Scott and I decided to buy our own 3D printer. That was probably our first significant capital expense, allowing us to make changes ourselves and iterate more efficiently.

Praveen: How much did it cost?

Dr. Steinberg: That one cost about $28,000, and it wasn't among the more sophisticated ones. It was suitable for our needs. There are now 3D printers that print with FDA-approved materials and can print commercially available medical devices. When we have something we want to change and go to market with immediately, we can 3D print it. It's a bit more expensive than doing molds, but it gets you to the marketplace faster while you work on getting the molds done for less expensive follow-on production.

Praveen: How interesting. I'm sure that the quality and cost of 3D printing technology have changed. What would that $28,000 printer cost now?

Dr. Steinberg: Without knowing the exact price, I'd guess it would be in the teens or even less than $10,000. Technology is advancing so rapidly that it's hard to keep track.

Praveen: Thanks for sharing that. I want to get to the money and the funding question. How did you and your team figure that out?

Dr. Steinberg: Credit cards. Both Scott and I were fortunate to be at a time in our careers when our kids were done with college, our mortgages were well on their way to being paid off, and we were relatively comfortable. So we could self-fund the initial aspects of the product, getting to the point where we thought it was worth pursuing further. At that point, we went to friends and family. I literally sent out an email to a large group of my Cornell friends, saying, "For those of you that have no interest in taking a risk, stop here and don't be embarrassed about not answering my email. For those interested in what we think is interesting technology and a potential investment opportunity, risky as it is, keep reading."

With that and the addition of Josh Cohn, our chief commercial officer, who had been the chief of marketing for endoscopy and ultrasound for Pentax, we made progress. Josh knew everyone in the space and was a great salesman, having sold me a million and a half dollars' worth of EUS equipment over the years. He reached out to many physicians who understood the potential significance of what we were doing, and they became investors. Our first round was friends and family, raising about $1.2 million, which went a long way.

Praveen: And subsequently, AGA also invested. Correct?

Dr. Steinberg: Right. After that, we won the AGA Shark Tank and represented both the AGA and the ASGE as innovation of the year. AGA, along with Varia, a VC company, formed an innovation fund and participated in our Series A. This was a huge help, both financially and as a stamp of approval. Fundraising is a real challenge because it's expensive to do the necessary testing and get through the FDA clearance process. The lack of clear pathways and funding capabilities for innovations is a significant hurdle.

Praveen: How much did they invest, or what is the total capital that you finally ended up raising?

Dr. Steinberg: To date, we've raised $11 million and are currently working on our Series B raise as we commercialize. We're in a better position now with FDA clearance and a pipeline of sites interested in our technology.

Praveen: I want to go back to your story. Why go through this trouble and pain of innovation at this point in your career? Why not just relax and continue whatever you were doing?

Dr. Steinberg: That's really interesting. I've never been able to leave anything on the table. Whether it was developing a therapeutic endoscopy program, and I think this is now my third or fourth one, I've always been driven by seeing something that could or should be done and feeling the need to do it. Sometimes, it's gotten me in trouble, and sometimes, it hasn't because people don't always want to hear about things you think are obvious opportunities. This one, fortunately, was under my control and Scott's control. As it evolved, it was slow at first. We got a beamformer, used the old technology from Scott's old company, and started getting some images. We had a phantom; the needle came out, hit the target, and we got excited. At that point, there was no turning back. It was too much of an opportunity, too much of a need in the healthcare space. And at that point, we were hooked. I've always been interested in things like that throughout my career. So this was just another one of those stages, although in a slightly different space.

Praveen: If everything turns out the way you're thinking, with commercialization and scaling, what kind of disruptive force would this device be in the industry?

Dr. Steinberg: That's an interesting question. We see this as a role for multiple technologies. Initially, our focus is on general endoscopic ultrasound (EUS). There are three layers of EUS. The first layer involves imaging only, using the results to draw conclusions or refer the patient to someone more experienced. The second level involves taking samples, such as fine needle aspirates or biopsies, for diagnostic material. This level is growing rapidly due to the need for samples and genetic analysis in personalized medicine. The third level includes interventional procedures like draining gallbladders, gastro-jejunostomies, and other extra-luminal activities.

Right now, we're most interested in the first two levels. In the US, fewer than 20 percent of general hospitals and fewer than 3 percent of surgery centers have EUS capability, which represents a huge unmet need. Outside the US, the need expands exponentially. The goal is to democratize EUS, making it a part of regular endoscopic exams. Why perform an endoscopy for gastrointestinal symptoms and only look at the surface when you could also image surrounding structures like the pancreas, gallbladder, and lymph nodes with ultrasound?

Legacy technologies don't lend themselves to this approach either, and we can make a transformational impact. There are 10 million upper endoscopies done annually in the US, and many patients end up needing further diagnostics like EUS after rounds of other treatments and scans. Making EUS a standard part of the initial exam could provide much more comprehensive information from the start. We call it "Echo-Endoscopy" – it would completely disrupt and change the landscape of endoscopy.

Praveen: What kind of advice do you have for other potential innovators out there? There may be somebody thinking, *I don't know how to go about it because I'm so busy in the endoscopy room. I have a family. I have kids. I'm busy.*

Dr. Steinberg: My journey was quite serendipitous. I was fortunate to train with some of the stars in the field. I've had the opportunity to watch therapeutic endoscopy evolve from the early days. I don't have a perfect answer to your question because we have so many things pulling at us that are really important—family, earning a living, having a satisfying professional life, and meeting all life's demands.

I was lucky that this idea blossomed at a time when my four kids were out of the house and my wife, who is the real star of the family, had her own successful career. We had the means to support this venture without mortgaging the house.

The truth is, it takes a lot of luck and circumstance. A great idea needs interest from others, often because it can make money. Unfortunately, our system doesn't always support great healthcare ideas that aren't profitable. I wish I had a more definitive answer, but it really depends on timing, luck, and having the right support.

Praveen: When you consider the future of gastroenterology, what concerns come to mind?

Dr. Steinberg: The challenges in gastroenterology are not unique to our field. Physician burnout is a common topic in media and publications. However, I believe burnout is not just doing the same task repetitively until you can't stand it anymore. It's more about wanting to do the tasks we are trained to do and get satisfaction from but being inhibited by forces outside our control.

A significant recent example is prior authorization. It's not just burnout; it's someone obstructing us from using our skills and education to make the best healthcare decisions for our patients. Factors like the EHR at 10:00 p.m. at night or the influence of hedge funds in healthcare, particularly in gastroenterology, are significant concerns.

These external factors hinder us from providing the best care. In gastroenterology, the hedge fund market has been particularly active and prominent. My biggest concern is ensuring we can continue to do what we know is best for our patients. Preserving our voice and our ability to provide optimal care amidst these market and external pressures is crucial.

Praveen: Thank you so much for reflecting on that very important question. It makes me reflective as well. It would be interesting to see how GI professionals will take to the future.

Dr. Steinberg: And so, from the GI docs' point of view, I'm fortunate in some ways. I do only therapeutic endoscopy, so I get to do the cool stuff and make a big impact on the patients I care for. I sit next to a group of extraordinarily well-trained general gastroenterologists and see how they are increasingly forced into throughput volume procedures. The idea of thinking outside the box is becoming more difficult and less available.

One of my thoughts about democratizing EUS and echo endoscopy is that it provides gastroenterologists with another tool and another way to expand their horizons. Hopefully, this will improve the healthcare we deliver and provide more professional satisfaction. As you say, I've been at this a long time, and the most satisfying moments were the changes that allowed me to do more, think about new things, and be more involved. Hopefully, democratizing EUS will offer similar opportunities for general gastroenterologists in the future.

Praveen: Thank you so much for joining me; this has been a great conversation.

Dr. Steinberg: My pleasure. And congratulations on your efforts. I think you bring a lot to the space.

SECTION: VOICES II - AI AND ROBOTICS

This section highlights the role of artificial intelligence and robotics in advancing gastroenterology, providing insights into the integration of these technologies into clinical practice.

HEALTHCARE EQUITY THROUGH AI IN GI: DR. JONATHAN NG OF ITERATIVE HEALTH

Dr. Jonathan Ng is the founder and CEO of Iterative Health. His background is in medicine, healthcare policy, and business. Iterative Health is a pioneer in applying artificial intelligence-based precision medicine to gastroenterology, aiming to establish a new standard of care for the detection and treatment of gastrointestinal diseases.

Themes: *healthcare disparity and equity, collaborative innovation, AI in GI, computer vision, self-driving in endoscopy*

Date: *October 2022*

Three Takeaways:

1. **Vision-Driven AI for GI Healthcare:** Dr. Jonathan Ng's company, Iterative Health, is focused on using AI to reduce healthcare disparities and improve patient outcomes in gastroenterology. Their AI technology aims to provide consistent and equitable information to all doctors, enhancing their ability to treat patients effectively.
2. **Integration of Computer Vision in GI:** Inspired by the use of computer vision in autonomous vehicles, Iterative Health leverages this technology to identify and analyze medical conditions in gastroenterology. This approach is designed to complement the expertise of gastroenterologists rather than replace them by providing additional data and insights for better decision-making.
3. **Future of AI in Gastroenterology:** Dr. Jonathan Ng anticipates gradual advancements in AI for gastroenterology, with significant improvements in polyp detection algorithms and personalized patient care over the next

five to ten years. The goal is to integrate multimodal data sources to offer tailored healthcare recommendations, moving away from generic guidelines toward more precise, risk-based assessments.

Praveen: Dr. Jonathan Ng, CEO and founder of Iterative Health (formerly Iterative Scopes), thank you so much for joining me on the Scope Forward show. I've been waiting to have this chat for many months now, so I'm glad we finally scheduled it. You've recently raised $150 million in Series B financing to accelerate the development of your core AI innovations. That's exciting and wonderful.

Dr. Jonathan Ng: Yeah, thanks for having me, Praveen. I'm excited to be here, and sorry it took us so long to get here.

Praveen: No, you've been busy raising the money that you did. First, congratulations to your team. It's very exciting not just for you as a company and startup but for the industry itself because it shows the interest that GI as a space is attracting. But Jon, I want to ask you first about the backstory here. You got the idea while you were on a trip in Cambodia, observing doctors trying to detect tumors. The idea germinated while you were doing your MBBS in Singapore. Please tell us all about it.

Dr. Ng: I'm happy to share that one. It's quite an interesting story. I got my start in healthcare pretty early on by chance. I happened to visit Cambodia. For context, I'm a first-generation college student with no doctors in my family. I was fortunate to receive mentorship from my friend's dad, who was a doctor. He invited me to Cambodia on a medical mission, and that was my first experience of healthcare outside of Singapore, which has a first-world healthcare system. It was shocking to see the state of healthcare in Cambodia, which was emerging from the Khmer Rouge regime.

After the regime, there were basically two doctors left in the country for a population of 12 million. The Khmer Rouge had killed off intellectuals to prevent uprisings. In Cambodia, I observed how kids were not surviving due to basic healthcare deficiencies. The under-five mortality rate was about 20 to 25 percent, which was unacceptable. I ended up spending about 14 years in Cambodia building various healthcare facilities, including pediatric cardiothoracic surgery units, neonatal wards, burn reconstructive units, and establishing pediatric hospitals in rural areas.

As part of this process, we were trying to train Cambodia's first generation of surgeons. I would bring in top tutors from Singapore and the US, from places like Brigham and Women's Hospital and Boston Children's Hospital, to share their knowledge. We frequently encountered a situation where my mentor would point out a massive tumor during surgery, and while I could see it and understand the details, the trainee surgeon couldn't even identify it. This inability to see the tumor meant the trainee couldn't treat it, leading to vastly different patient outcomes.

By my 12th or 13th year in Cambodia, I was doing a lot of regional work and felt frustrated by my inability to move the needle significantly. I felt like I was spinning in circles, constantly trying to make a difference. So, I decided to take some time off and go back to grad school. I had a couple of choices to make, but I really felt that I fit right in with the culture at MIT.

Praveen: Just to clarify, you were doing your MBBS and working in Cambodia at the same time?

Dr. Ng: Yeah, I started working in Cambodia when I was about 15 years old. I continued through military service, med school, and medical training, so I was involved in Cambodia throughout this period.

Praveen: Fascinating. Okay, please continue. So, now you're at MIT, and then what?

Dr. Ng: I came to MIT with an open mind, exploring various fields. However, my North Star was, "How do I use this opportunity to close in on outcome disparities?" One of the first technologies I encountered was computer vision in autonomous vehicles. This technology could identify cats, dogs, humans, and cars and even predict human movements. I thought, *This is fascinating. Why don't we have this in medicine?* It represented a major barrier to patients receiving equitable healthcare.

For the past 20 years, I've been working toward improving healthcare access and outcomes. I realized that traditional methods, like medical textbooks and one-on-one apprenticeship models in residency, weren't scalable. Conferences, while educational, aren't frequent enough and don't address fundamental, real-world questions effectively.

I saw the potential of using advanced technology in medicine. Although I'm not a gastroenterologist by training—I was training to be a surgeon—I was drawn to GI because of the potential applications of computer vision. GI is a visually based specialty, similar to surgery, but with cameras present in everyday procedures.

GI doctors are also great to work with. They don't take themselves too seriously, which makes the grind of innovation more enjoyable. Some specialties are more resistant to change, but GI seemed like a good fit.

We decided to focus on GI, aiming to close disparities in healthcare outcomes using this technology. I believe this vision accounts for much of our success, as many people see the potential as well.

Praveen: How does your AI solution differ from the ones already available in GI, like GI Genius™ from Medtronic or Satisfai Health? There are many startups waiting for FDA approval, and even Google has shown interest in this space with their work on computer vision and detecting polyps. So, what makes your solution different, and what similarities does it have with what's out there?

Dr. Ng: Great question. I won't claim to know every single solution out there. The biggest differentiator for us is the vision behind what we're doing and why we're doing it. We're focused on reducing health disparities and improving healthcare outcomes. We have a very practical, remote approach to applying AI in GI, which leads us into a more differentiated space. We're not just trying to sell more snares or SaaS products. For us, the fundamental reason is crucial and drives every decision we make at our company. I think while companies might look similar at the start, the everyday decisions that we make eventually turn us into very different companies.

Praveen: Somewhere, Jon, your message seems to have resonated with investors, especially those in digital health. You've raised a huge sum of money—over $150 million. I haven't seen such a large amount being raised in digital GI before. What attracted investors to your company?

Dr. Ng: Honestly, I wish I knew exactly. The breadth of our vision and the strong team we have in place played a big role. Since the early days, I've emphasized the importance of quality talent and a clear vision. These will define where each company goes. The GI market is big enough to support multiple AI companies. I always tell my team that our competition isn't really GI Genius™ or Satisfai; it's ourselves. We need to do what's best for our doctors and patients.

The quality of our people, the depth of our vision, and our ability to execute have all attracted significant investment.

Praveen: Did you plan to raise this amount of money at the end of Series A, or did it just happen?

Dr. Ng: It's a bit of both—serendipity and planning. It's been a huge privilege throughout this journey. I arrived in the US only four and a half to five years ago, spending two years in school. The past three years running this company have been incredible. Everyone has plans, but meeting the right people who understood our vision and hitting milestones helped. Being able to show that we could accelerate our progress with more funding certainly made a difference.

Praveen: So what is the holy grail for the company? What is the next big milestone from here?

Dr. Ng: The holy grail for us is to use AI and analytics to provide every single doctor with the information they need to treat their patients equally. Today, there's an unacceptable level of disparity in outcomes, whether we acknowledge it or not. I've worked in countries where we struggle to get basic care in place, let alone identify and treat complex conditions. By pushing this information into every endoscopic suite and minimizing the barriers for GIs to access and use it consistently, I hope we can equalize patient outcomes across the board.

Praveen: That's fantastic for patient outcomes, but some gastroenterologists might be thinking, "If a technology like this is going to do the job I do currently, what would I do?" How would you respond to that?

Dr. Ng: First, we're not trying to replace anybody. We're providing information so that doctors can do what they do best—treat their patients. All of us went into healthcare to help people. Why turn down knowledge and data that can help drive better outcomes? We're not taking over the mechanics of the procedure or replacing the pilot. We're making recommendations and providing a common identification of polyps and other features, leading to a consistent assessment of diseases. This helps create a common understanding of how patients should be treated. We're not taking away what GIs do today; we're enhancing their ability to do it better.

Praveen: But let's go back to that analogy of self-driving cars that actually prompted you in this direction. You may be familiar with the five levels of self-driving or autonomous vehicles. Somebody's given this classification, and I find it quite interesting. I even wrote about it in *Scope Forward*. My question is, what if that were to be applied to computer vision in the endoscopy room? If you were to apply levels one to five, with increasing levels of sophistication, where would you land when you apply computer vision to endoscopy?

Dr. Ng: Right now, what we're doing is very rudimentary. For healthcare, it just takes way too long to understand our impact on patients. Imagine this: it's already challenging to put autonomous cars on the roads, which are fairly standardized with straight paths, stop signs, and crosswalks. Now, think about anatomy—there are no such standardized markers. Everyone's anatomy is different, especially if they've had previous surgeries.

Given these complexities, it will take tens of years to achieve a totally autonomous situation in endoscopy. Until then, everything will focus on identifying patterns and providing support. We're still at a very basic level, perhaps akin to level one or two, where the technology assists but does not replace the human operator.

Praveen: That's interesting to know. I usually ask this question toward the end, but I want to ask it now. If you were to throw a stone and have it land five years from now, three years from now, and one year from now, where would AI in gastroenterology be? Let's start with the distant future—five years from now—and then move to three years and one year.

Dr. Ng: Five years is pretty short-term for us, considering the pace of change in healthcare. We might have some advanced polyp detection algorithms, specific features being detected, and perhaps some histology predictions used in clinical decision-making. That's where I see us in five years. I'm pretty conservative about how fast technology moves. Some visionaries might think we'll have self-driving scopes in five years, but I don't think so.

Praveen: Let's extend this horizon. If five years is short-term, what does increased sophistication look like? Let's push forward to ten years.

Dr. Ng: In ten years, being able to ingest multiple data sources that are becoming more available—like those from Guardant, Freenome, or liquid biopsy companies—will be key. We could risk-adjust and take a comprehensive view of a patient by combining various multimodal data sets. This would lead to personalized recommendations. Right now, colonoscopy screening recommendations are generic. People ask me what age they should start scoping, and I say as soon as you can afford it because I know people who've had stage four colon cancer at 35.

Having personalized, risk-based recommendations would be an incredible outcome. This approach could help us confidently and individually tailor healthcare measures, making it more efficient and cost-effective. Instead of broad guidelines, we could apply preventative measures and monitoring tailored to each patient's unique risk profile.

Praveen: Shouldn't we mandate using AI tools for endoscopies so that the ADR across the board reaches a certain baseline? Do you think that should happen, or is it likely to happen? Any comments?

Dr. Ng: I wish it happens—it's why we are here, and it's why we're present. I think it's a two-way street. We need to minimize the barrier to adoption. Currently, a lot of AI in its current form is either invisible or overly expensive for most GI practices to adopt. As a company, we have a responsibility to reduce or minimize this barrier to adoption and develop that ROI for our customers, and that's what we've been working toward time and again.

Praveen: Jon, it was great to have this conversation. It gave me a great perspective on how you're thinking and how the company is thinking. Is there anything else you want to share?

Dr. Ng: We're just really excited to keep innovating. We're always looking for good partners, so let us know—hit us up if you're keen to work with us or understand our products. We'll keep going.

ADDING A DIGITAL LAYER TO ENDOSCOPY: MATT SCHWARTZ OF VIRGO

Matt Schwartz is the founder and CEO of Virgo. He's a biomedical engineer and MedTech product expert. Virgo is building a digital layer for endoscopy and has amassed the largest endoscopic dataset in the world.

Themes: *video data innovation in endoscopy, machine learning, data ownership, clinical trials, pharmaceutical research*

Date: *June 2022*

Three Takeaways:

1. **From Surgical Specialties to GI Endoscopy:** Matt Schwartz transitioned from working on the Da Vinci robotic surgery system at Intuitive Surgical to founding Virgo. Initially focused on capturing video data in surgical specialties, Virgo pivoted to GI endoscopy after recognizing the immense procedure volume and potential in this field.
2. **Integration with Clinical Trials:** Virgo expanded its focus from merely recording endoscopy videos to supporting pharmaceutical clinical trials, particularly in the IBD space. Driven by feedback from principal investigators, Virgo's platform now aids in patient recruitment and central reading for trials.
3. **Adding a Digital Layer to GI Endoscopy:** Matt sees a significant opportunity to add a digital layer to GI and endoscopy. He believes that in the next five years, software solutions will emerge that go beyond report writing, enhancing patient care and clinical practice. Virgo aims

to be at the forefront of this transformation by continually refining its data capture and utilization processes.

Praveen: Before starting Virgo, Matt, you led product management teams in the minimally invasive spine surgery and robotic surgery fields. That's quite a switch from spine and robotic surgery to endoscopy video recording. Tell us, Matt, how did Virgo happen?

Matt Schwartz: It's an interesting story. I don't know if I would have predicted to end up in the GI space, but it's been a fun journey to get here. When I was at Intuitive Surgical, working on the Da Vinci robotic surgery system back in 2015, I became really interested and passionate about machine learning and computer vision. I felt there was this untapped potential for bringing machine learning systems into the world of video-based medical procedures. There's an incredible amount of video data being generated in these procedures that nobody was capturing, and if we could only start capturing it, we could build all the really compelling machine learning to go on top of it.

That's what led me to leave Intuitive and start Virgo. In the very earliest days of Virgo, we were primarily thinking about capturing video from surgical specialties. So, I spoke with orthopedic surgeons, neurosurgeons, and colorectal surgeons that I had worked with over the years. There was some interest in video capture, but it wasn't the intense interest you need for an early-stage startup. It wasn't the right product-market fit at that time.

I started doing some background research and recognized that GI endoscopy, from a procedure volume perspective, absolutely dwarfs all the surgical specialties. Maybe it doesn't get the same level of attention as surgery does, but from a volume perspective, there's just an incredible amount of healthcare going into GI endoscopy.

I connected with the dad of a childhood friend who's a gastroenterologist in my hometown of Indianapolis. He mentioned Dr. Doug Rex, one of the world experts in colonoscopy, who was right down the street from where I grew up. He suggested I speak with him because he might be interested in video capture. My cofounder Ian's dad was also seeing his gastroenterologist, Dr. David Cave, up at UMass. Between the two of them, they gave us a lot of encouragement that GI was a specialty to look at.

We ended up going to the AGA Tech Summit in April 2017 based on Dr. Cave's recommendation, and we met a bunch of forward-thinking gastroenterologists who thought what we were doing was really interesting. That, more than anything, pulled us into the gastroenterology space. It was an exercise in following the voice of the customer and seeing where things led us, more so than us having any bright ideas of our own and figuring that GI was the place to be.

Praveen: Very interesting. Virgo started an endoscopy video recording system. Now, you are helping with clinical trials. You seem to have pivoted in the last couple of years. Tell us about it.

Matt: Yeah, when we look at Virgo, our core competency is building automated video capture infrastructure in a cloud-based fashion. That's what the Virgo platform is built on—making it easy to capture high-quality endoscopy videos.

Initially, we sold this as a software solution to hospitals and health systems, and we built up a nice business.

However, we always believed the long-term value would come from the data being captured and building additional tools on top of that data. Just before COVID, we started exploring the pharmaceutical clinical trial space. This was driven by our key customers, many of whom were principal investigators (PIs) on pharmaceutical trials, especially in the IBD space. We repeatedly heard that the technology for facilitating IBD trials was lacking.

A big part of IBD trials is central reading, where you capture video data for study patients. Doctors were using challenging-to-set-up laptop systems, plugging everything in on the day of the case, and shipping the laptops around the world for central reading. PIs told us it would be great if we could improve central reading.

We realized the data we capture in endoscopy videos and standard-of-care encounters could help find the right patients for clinical trials. This pivot was driven by market need and the opportunities with pharmaceutical companies, which can be better early customers than hospitals due to their capital for new solutions.

Praveen: So, are your current customers pharma companies?

Matt: It's both pharma companies and healthcare providers. At the core of what we provide, it's crucial that we're helping gastroenterologists capture their endoscopy video data. Part of the value we add to clinical practice is enabling doctors to use Virgo for various purposes. They can build libraries and use videos for research projects, training residents and fellows, and quality improvement. Some doctors even share videos directly with patients to enhance engagement. We want doctors to gain value from Virgo daily.

Our other set of customers is pharmaceutical companies, where we help accelerate clinical trials. We're also working on trial site selection and optimization and starting to get into the central reading space. We see the platform becoming a central hub for not just IBD trials but other GI trials as well. It's like a two-sided marketplace now, with provider groups on one side and pharma companies on the other, and we're the technology solution in the middle, optimizing the integration between the two.

Praveen: On the physician side, what are they paying you for, and what is the business model there?

Matt: The fundamental business model is around video capture and access. We have an annual software fee, and we include the hardware as a loaner at no additional charge. With that annual subscription fee, the health system receives unlimited recording, storage, user accounts, and data access. We're very motivated to get doctors to record every procedure, not just some, which is a shift from past practices. Our model is fully unlimited.

We install devices in interventional and advanced endoscopy suites and can set up multiple devices to record endoscopy, fluoroscopy, SpyGlass, and endoscopic ultrasound simultaneously from the same room. Health systems interested in our clinical trial solutions can benefit from discount programs, which can significantly reduce the cost of Virgo. We monetize by working with pharmaceutical companies.

Praveen: Is there a business benefit for providers who record videos?

Matt: Yes, there are indirect business benefits, such as improving overall quality and enhancing training programs. One of the big business motivators is clinical research. Whether in private practice or academic groups, there's a strong motivation to have high-functioning research organizations. Efficient GI clinical research groups with limited overhead can become profit centers because pharma companies are highly motivated to find sites that produce high-quality patients for their trials.

We help clinical trial sites find more patients within their existing populations who are good candidates for research without adding additional overhead. Clinical trials tend to be very manual and labor-intensive, requiring skilled clinical research coordinators. Some groups, especially in private practice, have struggled to set up clinical research arms due to high overhead costs and low patient enrollment. We can be a free tool that helps drive a more profitable research center within their practice.

Praveen: How exactly does the system identify that a patient is suitable for an IBD or other trial? What is in the system that can identify this?

Matt: Great question. IBD trials are a great case study because they have specific endoscopic scoring criteria that a patient must meet to be eligible. Typically, finding a patient for an

IBD study involves the principal investigator (PI) within a practice evaluating their personal patient population. If they think a patient might be a good candidate, they approach them and suggest a colonoscopy to see if they meet the endoscopic criteria. However, many patients don't want to undergo an unnecessary colonoscopy and even those who do often don't meet the criteria. Screen failure rates on endoscopy are around 50 to 70 percent, which is frustrating for both patients and clinical research coordinators.

What we do with Virgo is capture every standard-of-care colonoscopy as part of the patient's normal journey. We've developed a machine learning tool called Auto IBD that creates a predictive score, which is not diagnostic but triaging in nature. The score indicates the likelihood of a patient being a good candidate for a trial. When a patient's score is high, we automatically alert the clinical research coordinator.

This approach is particularly impactful in large practices where many doctors have IBD patients, but only one is the PI. We help identify and refer potential candidates to the PI without extra work from other doctors. Most doctors, if they're not the PI, are not fully aware of ongoing research studies. We automate the identification process, using standard-of-care data as the first step to finding the right patients, thus flipping the traditional recruitment paradigm. Instead of hoping that a patient meets the criteria, we use existing endoscopy videos to streamline and improve patient recruitment for trials.

Praveen: How much video data is generated in GI on a given day, month, or year? I'm sure you've done some analysis on this.

Matt: Yeah, we've done some analysis. It's actually challenging to get accurate numbers on how many GI endoscopies are done in the US every year. We've seen estimates ranging from 15 million to upwards of 50 million annually. I think the right number is probably around 25 million GI endoscopies. Within our data set, we're now coming up on about 400,000 GI endoscopy videos that we've captured, and that number is growing almost exponentially.

How much data is being generated depends on the quality of the data you're saving and the sort of data compression you're using. We try to be intelligent about our data capture to minimize storage costs while preserving high-quality video data. For us, it's roughly one to one and a half gigabytes per hour of endoscope footage. So it's a lot of data, and there's still a lot of data going uncaptured today. One of our North Star goals is to change the standard of care. It's crazy that in 2022, with our ability to record unlimited 4K video on an iPhone, we aren't saving this crucial healthcare data. It's a bit crazy to me.

Praveen: 400,000 videos - that's a lot of videos. Who owns the data?

Matt: Our customers, the healthcare providers, own their data, and that's really important to us. There are a lot of incredible potential applications for this data, and we want our customers to be at the forefront of using their own data, whether for research, training, or other purposes. We also have academic partners leveraging their data captured in Virgo to build their own AI tools, which they might license. We want to support that.

With our customers owning their data, they grant us a license to use the data to provide it back to them in the platform and make improvements to our system. It's a typical SaaS agreement that allows us to furnish the services, ensure they can view their videos in the cloud, and make improvements over time with that data.

Praveen: Aren't doctors concerned that all their videos are being captured? The medical industry is highly regulated, and physicians are closely monitored. I suspect one of the concerns endoscopists have is that their data is now out in the open, and if something goes wrong, it's forever captured in the cloud.

Matt: It's a common objection. My dad, a spine surgeon, thought it was a terrible idea due to malpractice concerns. We understand that sentiment, but there's a shift happening.

We researched the liability implications and spoke with experts from malpractice insurance companies. About 60 percent of their costs go to legal fees and 40 percent to settlements. They prefer settling cases early to cut legal costs. Video evidence can be a great tool for protecting endoscopists. If a video shows the doctor performing standard care, it helps dismiss cases early. If it shows negligence, they'd rather quickly settle than incur high legal fees.

Some malpractice insurance companies even offer discounts to practices that record all their videos with Virgo. Dr. Doug Rex wrote an article explaining that in radiology, the entire MRI exam is captured, which he believes should be the standard for endoscopy. High-quality video recordings are more likely to protect doctors than harm them.

While this concern is common, there's a growing recognition that video recording is protective and the way forward.

Praveen: You've obviously heard the phrase "data is the new oil." The chokehold in the AI space is no longer the algorithm. It used to be about creating better AI algorithms, but now, many are open-source and available for free. The real chokehold is data—clean, usable data. There's a huge opportunity for endoscopists and providers to benefit through partnerships with companies like yours that can make this data usable. It seems like this will explode exponentially.

Matt: We realized early on that the algorithm wasn't the secret sauce. With so much open-source technology available, even non-experts can build interesting machine-learning tools. The real differentiator is the data—how much you have, the quality, and how you can refine it. One overlooked aspect is the freshness of the data. The best AI technologies, like self-driving cars or Google Search, constantly refresh their data streams, allowing them to refine algorithms in real-time.

For us, it was crucial to have a continuous flow of data rather than just a static set. We could have captured 10,000 videos and stopped, but we wanted ongoing data input. Another key aspect is figuring out the right business model for AI. I'm skeptical about the future of clinical decision support from a business perspective, particularly around reimbursement. Cracking the business model and distribution will be critical for AI's success.

Praveen: Let's switch gears to the business aspects of your company. Congratulations to you and your team for the great fundraiser. What can you share with us about this fundraiser?

Matt: This is our Series A raise. We raised $8 million, which we're incredibly excited about. Our lead investor group is FCA Venture Partners from Nashville, which is great since my cofounder and I both went to Vanderbilt. They focus exclusively on healthcare and have deep networks with healthcare provider groups. They've also invested in digital health companies that provide software services to pharmaceutical partners, so they've been an amazing partner.

We also have a strategic investor coming on board, making their first investment out of their corporate investment fund. I can't announce who it is yet, but we will soon. We're thrilled to have them as a partner, not just from an investment perspective but for the long-term strategic relationship that will drive our growth forward.

With the Series A, our focus is on building out the team. We've already added several people this year, and we plan to double our team size by the end of the year. We're adding people to sales and customer support, as well as building out our engineering team.

Praveen: You have a bird's eye view of the GI space, I must say. You're connected to so many different practices and providers, and you're seeing the industry from the Pharma side as well. What is the future of GI from your lens as an innovator and engineer?

Matt: Take everything I say with a grain of salt—I'm certainly biased—but I still feel there's a huge opportunity to add a digital layer to GI and Endoscopy. There's so much potential for software solutions to help out, and we're still in the early stages. We're friendly with several digital health companies like Sonar MD and Oshi Health, and I find what they're doing incredibly compelling, especially in bringing the patient experience into the fold with software.

In the next five years, I think we'll see a digital layer emerge on top of Endoscopy that goes beyond just report writing. EHRs are the current digital layer of healthcare, but I've always felt that calling them electronic health records is a misnomer—they're really electronic billing records designed to facilitate optimal billing in healthcare. There's still an opportunity to build software that creates a proper health record, a clinical record that's viable for day-to-day clinical practice.

Praveen: Matt, it's been a fantastic conversation. Thank you for sharing your perspective and being open about everything I asked. Is there anything else you want to share?

Matt: This has been fantastic. Thanks so much for having me. I have always enjoyed your interviews, and you do an awesome job of bringing together different minds from the industry. I especially appreciate your willingness to put out controversial opinions. I look forward to more of your interviews—keep it up.

THE ELEPHANT IN THE ROOM IS REIMBURSEMENT FOR AI ADOPTION: DR. MICHAEL BYRNE OF SATISFAI HEALTH

Dr. Michael Byrne is the founder, Chairman, and CMO of Satisfai Health. He's a practicing physician and clinical innovator specializing in interventional endoscopy and gastrointestinal diseases. Satisfai Health specializes in AI-driven medical solutions for gastroenterology, providing real-time analysis and decision-making intelligence to improve patient outcomes and clinical efficiency.

Themes: *AI in GI, polyp detection, optical biopsy, personalized medicine, reimbursement for AI, regulatory update, AI as hybrid intelligence*

Date: *September 2023*

Three Takeaways:

1. **Innovative AI Applications in GI:** Dr. Byrne emphasizes the significant advancements Satisfai Health is making in AI applications for gastroenterology. They are developing tools for polyp detection, quality measures during colonoscopy, and automating reads for clinical trials in Crohn's and Colitis. Their work in the Barrett's esophagus space, focusing on early cancer detection and guiding biopsies, illustrates the innovative use of AI in improving diagnostic accuracy and patient outcomes.
2. **Challenges and Opportunities in AI Adoption:** The adoption of AI in gastroenterology faces hurdles, primarily due to concerns about efficacy, lack of education, and reimbursement issues. Dr. Byrne highlights that demonstrating cost-effectiveness and utility in real-world

practice is crucial for wider acceptance. While there are barriers, the potential benefits of AI, such as enhanced diagnostic accuracy and workflow efficiency, are driving the field forward.
3. **Future of AI and Personalized Medicine in GI:** Dr. Byrne envisions a future where AI tools act as expert assistants to physicians, enhancing their capabilities and ensuring high-quality care. He believes in the potential of AI to integrate seamlessly into clinical practice, providing personalized and precise medical care. This convergence of AI with other technological advancements, like liquid biopsies and stool testing, will transform the landscape of gastroenterology, making procedures more targeted and efficient.

Praveen: Dr. Michael Byrne, give us an update on Satisfai Health. Where are things with the company?

Dr. Michael Byrne: It's been a very busy 12 to 18 months. Most of our business in the last couple of years has been good. We've grown the company quite significantly, with over 25 people now. We have been in acquisition mode, as you may have read last year, with the acquisition of Docbot. We've continued to build a suite of solutions that we aim to knit together. There's a continuum of solutions in the clinical practice space and in the clinical trial space, particularly in relation to IBD, such as Crohn's and Colitis. We've engaged some industry in the pharmaceutical world for some projects I can't go into too much detail about. What I can comment on is that we forged a partnership with Alimentiv (formerly Robarts). It's a GI-specific IBD niche clinical research organization (CRO) based out of London, Ontario. We're delighted to work with that group.

Praveen: Congratulations. What does Satisfai Health offer in terms of products and solutions? What is in the portfolio right now?

Dr. Byrne: Most of our tools are still going through the regulatory approval process and/or clinical trials as required. We are at a very advanced stage in the polyp space, particularly in colon polyp detection (CADx) and colon polyp differentiation (CADe). We're also focusing on quality measures during colonoscopies, such as bowel prep automation, recognition of landmarks like the cecum, and timing of withdrawal.

In the Crohn's and Colitis area, we're automating the reading of videos collected when patients are potentially entering clinical trials. Instead of always needing a central reader, we've developed an AI tool that can augment, abbreviate, or potentially replace the reader over time with the right approvals.

For the upper GI tract, we're doing a lot of work in the Barrett's esophagus space, particularly in early cancer detection and finding dysplasia. We aim to move away from non-targeted biopsies and direct physicians to areas of concern. This is a very exciting and much-needed space. We recently forged a partnership with Oxford University for some Barrett's work that their group has been doing, now under the Satisfai umbrella. We look forward to collating all of these tools together into a comprehensive Barrett's tool.

Lastly, we have a partnership with the International Bowel Ultrasound Group (IBUS) to develop AI solutions for intestinal ultrasound, specifically handheld transabdominal point-of-care ultrasound for Crohn's and Colitis. This is quickly gaining traction among GI doctors globally.

Praveen: What are the takeaways from your new book, AI in Clinical Medicine?

Dr. Byrne: One of them might be don't do another book. It is a huge amount of work, even if you dedicate your entire time to it. That said, my other positive takeaway was the huge network that has been developed. I now have friends and colleagues in the medical AI space across the globe and across various specialties. There's lots of overlap, and there are many things we can learn from people in radiology, surgery, and other areas.

The idea here was not to create a high-level AI textbook—there are plenty of those out there. This book aims to bring the concept of AI and medicine to the day-to-day healthcare professional. We cover the basics of AI, the building blocks, how some of the algorithms are put together, ethics, safety, and risk at the beginning. Then, we have a fairly comprehensive middle section covering most fields in clinical practice. Toward the end, we focus on how to incorporate these technologies into real practice, looking at reimbursement and moving beyond aspirational AI to truly day-to-day AI in medicine.

Praveen: Congratulations on the book as well, Mike. If you were to paint the landscape of AI in gastroenterology, what does that look like?

Dr. Byrne: Many things have moved forward very well, but you could also argue that we've been stuck a little bit. If you look at the mass adoption of AI or the availability of multiple different types of tools in GI in North America, it's pretty much limited to polyp detection rather than other technologies that are being licensed in Europe for biopsy in polyps, quality assurance, Barrett's Esophagus, etc. It's a little surprising. Had you asked me three years ago if we'd have more than polyp detection approved and in clinical use in North America, I would have said absolutely. But we're not there yet.

However, a significant shift is happening, hopefully for everyone's benefit. One of the barriers we've faced is the need for robust evidence. Standalone testing showing accuracy is great, and many groups, including ours, have shown that AI tools perform well on specific datasets. People want to see randomized controlled trials (RCTs), which are very important for establishing evidence. However, RCT evidence alone is probably not enough to convince many people to adopt AI and push for clinical AI. What we now need are post-marketing or post-product placement studies showing cost-effectiveness and utility.

Those studies are starting to come through. Some of my colleagues and competitors are doing great work looking at the utility of tools like CADe polyp detection in day-to-day practice. Hopefully, that's where the value will be seen, and we will move past this slight sticking point in terms of adoption.

Praveen: What's been the biggest bottleneck for adoption? It's more than the technology.

Dr. Byrne: It is. It's concerns about true efficacy, whether that's real or not. I won't quote names, but I've spoken to a number of colleagues in GI globally when I go to meetings, and some of them point blank say they don't think they need it. Sometimes, the internal voice is, "Well, if your adenoma detection rate is 80 percent, then probably you don't. But if it's like 35 or 40, you can always get better, and technology will help you."

Some of it is just a lack of education, experience, or insight. That's why I took on that book project: to try to disseminate that message a bit more. The biggest one, though, is the elephant in the room: reimbursement. Who's going to pay for this? These technologies, as you know from your many conversations, are not cheap to develop, put through trials, and navigate regulatory approval before even getting to market and distribution. It's not for the faint-hearted. If people don't see a reimbursement path, that's a significant barrier. I think those are the key ones.

Praveen: So let's talk about the elephant in the room: reimbursements. One of the biggest questions about AI has nothing to do with AI. It goes more like, who's going to pay for it? So who's going to pay for it? What do you know about the direction in which insurance reimbursements are going? Are alternative payment models emerging?

Dr. Byrne: Yeah, I mean, it's very important to address. A lot of people joke and say, "Oh, nobody's going to pay for it; it's just going to be an expected tool." Well, maybe if you're a large multinational industry leader, you can do that. But in general, for the whole range of tools in GI or in medicine, there needs to be commercial viability for products to survive and be useful.

Who's going to pay? We are seeing some emergence of CPT codes in the US, for example. There have been two or three obvious ones in the last few years, like Viz AI and IDX for diabetic retinopathy, and other examples coming through of CPT or new technology add-on payment type codes in radiology.

So, there's every reason to believe that those will come to GI. Remember, radiology started this AI journey way before we did. Many people say that GI has actually taken the bull by the horns. I read something a year or two ago, possibly on your blog, that the biggest number of RCTs in AI in the medical space was actually in GI endoscopy. Some of those fee codes will come to GI, hopefully in the not-too-distant future.

Or is it value-based payment? Do we need to show cost-effectiveness in the market, show that we save time, buy back physician time, improve accuracy, and improve outcomes? Maybe part of the payment back is from, for example, in the US, from the HMO or the third-party insurer, depending on the situation. There are several examples of where payments can go.

Having said that, I was just at a meeting in Washington, DC, over the weekend. There was a very interesting session in the afternoon led by the faculty there on reimbursement barriers and opportunities in GI. Lots of, "Well, it could be this, and it could be that, and it could be the other," but still, in the end, most of us left a little unsure as to how and maybe when reimbursement comes in one or many guises.

I can confidently say that somebody somewhere will pay for this. The healthcare system, in general, will see the utility and the benefit of AI. We're just at the beginning of seeing the advantages. Of course, there are downsides, and we can talk about that if you like, but the benefits of AI in medicine, in GI, and in endoscopy are literally huge. The ability to drive toward personalized medicine, precision endoscopy, or whatever you want to call it, is very significant. As soon as the value of that is even sniffed by the paying bodies, the floodgates will open. That's my fairly strong prediction.

There are too many people staying in this space. If it was a fad or something that clearly is not going to be adopted by healthcare professionals, many people would raise the white flag and move on to the next cool thing in technology. And they're not because we all see it coming.

Praveen: The game has just about begun, so we're seeing all these early signals. Talking again about reimbursement, when I interviewed you for my book *Scope Forward*, you made an interesting prediction. You said that the real game will begin when insurance companies mandate a certain quality measure for things like ADRs. For example, if they say, "We'll only pay you this much if you meet a certain standard for adenoma detection rate," then it will change the game. Where are we with that type of thinking? Do you think insurance companies are as far ahead as well?

Dr. Byrne: I know they're thinking about it, but maybe not in the polyp space yet. It behooves the societies—the AGA, the ASGE, the ACG, the Multi-Society Task Force—to push for reimbursement for tools that help with polyp detection. If they believe in it, which I know they do, third-party payers and insurance companies will follow. We might be focusing on the US market here, but these principles can apply elsewhere, even though the systems differ.

Take the Crohn's and Colitis space, for example. If you're on a particular biologic and want to switch, it's currently a somewhat subjective decision by the physician based on symptoms, maybe some stool biomarkers, and often what we see during a colonoscopy. These are good but not great. Machines can see disease activity at a granular level more consistently than humans, who can fatigue. I believe insurers are thinking about holding doctors to higher standards of diagnostic accuracy, especially when it involves expensive drug treatments.

I thought we might already be there with polyp CADx or optical biopsy, given several modeling studies suggest you can discard small polyps after an optical biopsy. However, there's resistance despite promising AI technologies. Some of this is turf protection, which encroaches on pathology practices. These real-world factors complicate the adoption of new standards. It's not just about doing the right thing; it's also about whose space is being encroached upon. That's the nature of human progression.

Praveen: Let's talk about the cost and reimbursement of AI technologies. While AI is expensive to build, it tends to get cheaper over time, following Moore's Law. So, while initial costs are high, the value of these technologies can become immense as they become cheaper to use. How soon do you think we'll see AI in gastroenterology, like polyp detection, become more affordable and widespread, similar to how generative AI is creating visuals and videos now?

Dr. Byrne: You make some great points. The initial costs for developing these tools—training, commercial viability, trials, regulatory approval, and maintaining safety—are indeed high. But you're right; costs can come down over time. Look at how the price of VHS players plummeted from hundreds of pounds to just a few dollars.

Even as technology becomes cheaper to develop, there's a continuous push for improvement. AI tools today assist with detection and decision-making, but their future lies in predictive and personalized medicine. The granular insights

AI can provide, predicting patient outcomes based on various factors, are incredibly valuable. However, getting to true precision medicine, where treatments are tailored to each individual, is a significant leap that will require substantial time, investment, and money.

The value these tools offer justifies their cost. As they evolve and become more integrated into healthcare, their worth in delivering personalized medicine will be immense.

Praveen: I want your thoughts on convergence. AI is becoming more advanced and affordable, and we see similar progress in digital biology with liquid biopsies, stool DNA, and RNA testing. These advancements are targeting multiple cancers, not just one. Do you see these fields converging in the workflow of the average gastroenterologist? How do you visualize that future where AI and digital biology are routinely integrated?

Dr. Byrne: I recently gave a talk at the Mayo Clinic on incorporating AI into practice and some futuristic thinking. My advisor, Dr. James East, often talks about high-performance screening, which combines accurate liquid biopsies or stool biomarker assays, AI-driven imaging, and a dedicated screening program. Together, these can help plan most colonoscopies as therapeutic procedures.

Currently, no technology can remove polyps automatically; endoscopists still need to physically remove them. If we use high-performing biomarkers to screen and validate the signals with AI-assisted imaging, it ensures that polyps aren't missed. For example, if a liquid biopsy indicates polyps but the endoscopist misses them, it seems like the biopsy was wrong when it's actually a human error. Marrying AI with downstream imaging after an initial positive signal can improve accuracy and reduce unnecessary endoscopies.

Many endoscopies are currently unnecessary and waste resources. By fine-tuning the screening process with advanced biomarkers and AI, we can ensure procedures are therapeutic and necessary, reducing unnecessary interventions and improving patient outcomes.

Praveen: Awesome, and it's going to be interesting to see how that shapes up. One important thing I want to get out of the way is the regulatory landscape surrounding AI. What's the latest, especially for GI?

Dr. Byrne: Right now, in the US, the only approved AI technology in GI is CADe polyp detection. In Europe, there are a few more, like optical biopsy, Barrett's, bowel prep, and some QA measures. The FDA has appropriate barriers to entry, which is crucial to avoid chaos and ensure safety. However, regulatory bodies need to adapt to newer concepts, like optical biopsy with AI. The evidence shows that AI can perform these tasks well, so I expect the FDA and European CE processes to evolve. Despite being somewhat stuck in the past few years, I believe we'll see significant changes in the next twelve months.

Praveen: Mike, as we begin to wrap up our conversation, I have two final questions. First, what's your advice to private practice GIs in the US?

Dr. Byrne: Your question points to the idea that AI adoption mainly exists in academic settings for now. It's seen as a cool tool providing more insight, while private practice focuses on quick turnover and high volume. I'm not saying it's low quality; it's just very business oriented. Private practice shouldn't worry about AI tools diminishing income or changing their practice negatively. The true value of AI will be seen when we have a comprehensive integration of various tools into the patient flow. AI can help with quality assurance during procedures, identify lesions, assist with optical biopsies, manage bowel prep and withdrawal times, and generate reports. This integration will save time, reduce administrative burdens, and ultimately benefit private practices by streamlining processes and improving efficiency.

Praveen: Excellent. Second, if you have to present the future of gastroenterology from your lens, what does it look like?

Dr. Byrne: I see the future of gastroenterology as one where you essentially have an expert on your shoulder. Imagine having a global expert in endoscopic surgery for early cancers in the stomach guiding you. This expertise usually requires extensive training and constant practice to maintain. Most practitioners don't have that level of expertise. That's why when we miss something, it's a problem. But if you have an AI tool acting as that expert on your shoulder, it enhances your capabilities. It's about physicians and machines working in perfect harmony. There's a popular saying that AI won't replace doctors, but doctors who use AI will replace those who don't. I agree with that, but I see it as hybrid intelligence—physicians and AI working together to improve outcomes. It's an exciting and transformative time for the field, and I'm very optimistic about the future. However, I don't think I'm ready to write another book just yet!

BRINGING AI TO THE ENDOSCOPY ROOM THROUGH GI GENIUS: DR. ANDREA CHERUBINI OF COSMO IMD

Dr. Andrea Cherubini is the senior vice president for science, AI, and data at Cosmo IMD, the creator of GI Genius™. His background is in academia and MedTech, specializing in medical imaging and computer-aided diagnosis. GI Genius™ is the first-to-market, computer-aided polyp detection system powered by artificial intelligence (AI).

Themes: *GI Genius™, real-time AI in GI, polyp detection, Open AI platforms, research validation of AI, physician resistance to AI, global reach*

Date: *March 2024*

Three Takeaways:

1. **Comprehensive AI Integration in Gastroenterology:** Andrea highlighted that GI Genius™ began with polyp detection but has rapidly evolved to offer a range of AI applications for the entire endoscopy room. This shows the potential for AI to transform various aspects of medical procedures, ensuring higher quality and more efficient diagnostics and treatments.
2. **Collaboration and Open Platforms:** The AI Access initiative, launched with Medtronic and Nvidia, invites other startups to develop their applications on the GI Genius™ platform. This open platform approach accelerates innovation, making advanced AI solutions accessible to a broader range of medical practitioners and ultimately benefiting patient care.
3. **Future Vision and Continuous Innovation:** Andrea emphasized the importance of continuous innovation

and thinking ahead. While current AI applications are already improving the quality of procedures, future advancements could include predictive analytics tailored to individual patient histories. This forward-looking perspective is crucial for staying relevant and advancing the field of gastroenterology.

Praveen: Andrea, I want to start from the beginning. How did Cosmo IMD, as a division, get formed leading up to GI Genius™, the first commercially available AI device in GI?

Andrea Cherubini: It was like the perfect storm. I come from academia and have been interested in medical imaging and computer diagnosis for most of my career. My partners, Nhan Ngo Dinh and Giulio Evangelisti, and I had a company specializing in medical imaging. Among our clients was Cosmo Pharmaceuticals, a company specializing in GI.

The perfect storm happened because Cosmo had the opportunity to conduct a clinical study with the quality of a typical pharma clinical study, allowing us to gather many colonoscopy videos. In 2017, recording video colonoscopies wasn't common. We saw an opportunity to develop a solution for polyp detection to reduce the number of missed lesions in colonoscopies, even before AI was the buzzword it is today.

Cosmo Pharmaceuticals was enthusiastic about the idea, so we embarked on the project and quickly prototyped what would become GI Genius™. We started with a high-quality dataset, the best AI engineers, and a solid understanding of medical imaging and clinical studies. Medtronic was very interested in AI, and our agreement with them was like the perfect storm. That's how everything started.

Praveen: Superb. Can you bring us up to speed with the developments and updates with GI Genius™ since its launch, first in Europe and now in the US as well?

Andrea: Many might think GI Genius™ is just for polyp detection, but for us, polyp detection was just the beginning. Our aim has always been to develop a solution that brings AI to the entire endoscopy room in real time and beyond. We were the first to clear a polyp detection system through the FDA's De Novo process, which we consider a per-frame AI, meaning it outputs a prediction on each frame, answering the question, "Is there a polyp in this frame?"

In our pipeline, we intended to move to a per-object AI, where the dimension of time comes into play. For example, we have a feature called characterization, cleared in Europe and other countries (but not yet in the US), which anticipates the potential histology of each lesion. This is a per-object AI because it collects several opinions across all frames where a polyp is present to output a comprehensive prediction.

We are moving even further. Today, we have FDA clearance for a new AI version that gives predictions at a per-patient level. It analyzes the entire procedure and outputs critical quality parameters like bowel cleanliness, cecal intubation, withdrawal time, and procedure time. These parameters are linked to the quality of the colonoscopy, which is itself linked to the ADR and other key performance indicators.

In our pipeline, we have even more impressive AI. The power of AI today is incredible, and the next generations of GI Genius™ will keep improving with the rapid advancements in this field.

Praveen: Since your launch, there have been several AI approaches in gastroenterology. How does GI Genius™ today differ from the other products?

Andrea: Thank you for this question. Since the beginning, the GI Genius™ module—the hardware part that comes into the endoscopy room—was designed with the concept of hosting several AI engines. The hardware used for inference is really powerful. It was designed to be very powerful from the start, even too powerful just for polyp detection, because we had a broader plan in mind.

First, we believe that real-time AI in the endoscopy room is crucial for safety reasons. Having the ability to make an AI prediction in the endoscopy room minimizes lag time. This is already a differentiator compared to some other companies that may provide the output on a second screen or use the cloud.

We also recognize another point you mentioned: there are many startups, which is true because there are many unsolved clinical needs in this space. We believe the only way to bring all this potential AI to patients, which is our goal, is through collaboration. That's why, last year, we launched the AI Access initiative together with Medtronic and Nvidia. This initiative is a marketplace designed to solve the time-to-market problem for startups.

In simple terms, it allows startups to use the GI Genius™ hardware as a vector to bring their solutions to patients. You can think of the GI Genius™ hardware as a smart TV. One of the applications will always be the one we developed for colonoscopy at Cosmo IMD, which will be called Colon Pro from this year. However, there is room for other potential applications in the colon, the upper GI tract, and endoscopy in general.

Praveen: I was there at the GI Genius™ summit last year when this announcement was made with Nvidia, Medtronic, and Cosmo. I thought it was very innovative to come up with a platform inviting other startups to develop on your platform. What progress has been made since that point? What kind of applications are coming out right now?

Andrea: I cannot disclose specific applications or the companies that have approached us, but there has been significant interest from many companies in this field. Some are in the exploratory discussion phase, while others are much further along. These advanced companies are deep into the workflow of bringing their applications to the GI Genius™ platform and are already in the regulatory stage to approve their software as a medical device. Once they complete this regulatory process, we can deploy their applications on the hardware distributed by Medtronic.

Praveen: Very interesting, Andrea. I'm curious about the numbers. What is the scale of adoption here?

Andrea: As you can imagine, there aren't hundreds of unmet needs, but there are certainly tens of them, and each has multiple startups with unique approaches to solving these issues. Our goal with AI Access is to create an open platform. We believe that any company or partner wishing to bring their application to AI Access can do so, provided they meet the high-quality standards required both from a regulatory standpoint and because the AI will be hosted on

Medtronic-branded hardware. So, we are open to anybody who meets these criteria.

Praveen: If a GI innovator wishes to make something happen on the GI Genius™ platform, how do they go about it?

Andrea: On our website, cosmoimd.com, there's an Innovation Center page where anyone with an idea, whether they're from a well-established company or a single innovator, can apply. We've structured a detailed workflow for participating in this initiative. It's important to note that while we encourage everyone to apply, companies already in an advanced stage of product development will find the process easier due to their familiarity with industry requirements. However, we still welcome individual inventors. In the interest of transparency, we've shared a public dataset of colonoscopy videos available under the name RealColon. This allows anyone interested in developing AI solutions for colonoscopy to start testing their ideas and potentially turn them into products that can save lives in GI. You can find all this information on our website.

Praveen: Excellent. A lot of research has come out since the launch of GI Genius™ in the US, and these publications repeatedly demonstrate that AI is indeed useful. What have you learned after all these publications?

Andrea: We've learned a ton. It's been a privilege to see the scientific community engage in this discussion. I'm really proud that in just a few years—since the De Novo process cleared GI Genius™ in 2021, which seems like ages ago but is just three years—we've already had more than 30 publications, including eight randomized control trials of this device. That's incredible.

What does this mean? First, it means that GI Genius™ is one of the most tested AI platforms in medicine in general. It also means we've learned several important things about using AI in the field and in the real world. For example, we've observed that physicians' acceptance of AI significantly modulates the amount of benefit they gain from it. Some physicians are natural at using AI and see their ADR (adenoma detection rate) improve incredibly, while others, perhaps with a lower level of acceptance, see fewer improvements.

This presents a challenge for us and the scientific community: how can we ensure everyone experiences the positive effects of AI? We need to understand the factors that influence this variation.

Another crucial insight we've gained is about human-AI interaction. In the AI characterization process, where the AI outputs a prediction around a single polyp, we've asked ourselves whether physicians would follow the AI's suggestions blindly, refuse them completely, or something in between. What we discovered is incredible: in the right conditions, a human-AI hybrid team can achieve higher accuracy than either the human or AI alone.

The key is that the AI must communicate the boundaries of its predictions very clearly to the physician. This allows the physician to leverage the best of AI while relying on their own judgment whenever they feel the AI is uncertain. This discovery is fascinating and suggests we may need to rethink how we train physicians in the future.

Praveen: What is the biggest resistance from physicians that you see?

Andrea: Engaging in discussions with several physicians using AI, I've noticed a distinct generational divide. Younger physicians, who are more tech-savvy, have been exposed to AI from the beginning and see it as an integral part of their practice. They can't even imagine not using AI.

However, there's a generation caught in between who have experienced both pre-AI and AI-enhanced practices. Within this group, there are several factors at play—acceptance of technology, reluctance to change established routines, or other forms of resistance. While we don't have a clear answer yet, it's crucial to understand and address their concerns.

I empathize with this group. It reminds me of when I first had a car with a rear radar to assist in parking. Initially, the beeping sounds annoyed me, but now I can't imagine parking without this aid. Similarly, the challenge is to help these physicians see the long-term benefits of AI despite their initial resistance.

Praveen: AI development has accelerated across all domains. How difficult would it be for one of these increasingly powerful AIs, like those from OpenAI or Stability AI, to develop an algorithm that competes with GI Genius™? Do you worry about this?

Andrea: First of all, I welcome technology. I believe that enabling tools should be embraced positively, and this applies to medicine just like any other field. So, no, I'm not worried. We always think ahead. As the senior vice president of science, AI, and data, my job is to envision the products that will emerge in three to four years. Today, my team and I are discussing ideas that seem like science fiction, but they are possible.

What I'm seeing is that AI excels at automating tasks and handling repetitive and time-consuming activities. This doesn't just apply in real time but extends to all aspects of daily routines, freeing up time for more important tasks. The flip side, which sounds promising, is that humans will always be in the loop, especially in regulated industries like healthcare. This is crucial because AI is being used on patients, so it must follow regulatory processes.

Ultimately, what will be left to physicians will be the most challenging parts of their job. It's similar to what happens with airplane pilots today; 99 percent of the time, they monitor while the computer does everything, but they must be there when things go wrong. In healthcare, I see a trend where AI will handle the mundane tasks, but human expertise will always be needed for the complex decisions.

Praveen: GI Genius™ has thousands of installations across the world, with the majority in the US. What do you see happening over the next three to five years?

Andrea: First of all, I hope the success of GI Genius™ with our partner, Medtronic, continues to build momentum, maintaining its position as the most powerful and appreciated AI platform in this field. I also hope that our AI Access initiative becomes a gateway for many other companies to deliver their applications through this platform, not only for colonoscopy but for endoscopy in general and even beyond. There are numerous unmet clinical needs, not just in video detection but also in optimizing the overall quality of procedures in the operating room. With the latest version of our software cleared by the FDA, we are already addressing quality issues during procedures, but this can be expanded even further beyond GI.

Praveen: My final question, Andrea, and I ask this in every interview. What is the future of gastroenterology from your perspective?

Andrea: I believe this technology is an enabler of several crucial issues. One major issue is quality. Increasing the quality of procedures benefits patients significantly. Another critical aspect is access to screening. By enabling complex or difficult procedures to be more accessible, even to less experienced physicians, we can ensure that patients, including those in underserved communities, have access to high-quality procedures.

Finally, there are potential applications of AI today that could make what seems like science fiction a reality—such as predicting when a lesion will appear or tailoring procedures based on a patient's medical history. These possibilities are technologically feasible today, and I look forward to seeing them implemented in the near future.

Praveen: Great. And I'm looking forward to that future. Are there any final comments that you want to leave us with?

Andrea: Well, I would like to thank you for giving me the chance to talk about these topics that are very close to my heart. I feel very fortunate to have had the opportunity to bring AI into the real world and to share this experience. It's been a privilege, and sharing this privilege with you is very important to me. So, thank you.

FROM SELF-DRIVING CARS TO SELF-DRIVING ENDOSCOPES: SAURABH JEJURIKAR OF ENDOVISION AI

Saurabh Jejurikar is the founder and CEO of Endovision AI. His background is in software development, fintech, and travel-tech. Endovision improves the diagnostic accuracy and efficiency of upper GI endoscopists through real-time AI assistance, leveraging deep learning and computational modeling.

Themes: *cross-disciplinary innovation, self-driving vehicles, AI in upper GI, global healthcare collaboration, quality control via AI*

Date: *November 2023*

Three Takeaways:

1. **Innovation through Cross-Disciplinary Expertise:** Saurabh and his cofounder Deepanshu leveraged their backgrounds in AI and autonomous vehicles to address challenges in gastroenterology, particularly focusing on reducing blind spots during endoscopy procedures.
2. **Global Collaboration and Reach:** Despite being based in Hong Kong, Saurabh's company has established a global presence with a development team in India, collaborations with key opinion leaders in multiple countries, and a focus on the US and European markets. This demonstrates the potential for healthcare innovation to transcend geographical boundaries.
3. **Empowering Junior Endoscopists:** The product aims to bridge the skill gap between junior and experienced endoscopists by using AI to enhance diagnostic accuracy

and coverage. This can democratize access to high-quality endoscopic procedures and improve overall patient outcomes.

Praveen: Saurabh, what intrigued me about your company was this unique angle. Your cofounder, Deepanshu, comes from a self-driving or autonomous vehicle background. You mentioned that your father was a gastroenterologist. How did you guys make the connection from self-driving technology to endoscopy?

Saurabh: Yes, that's quite an interesting story. As you mentioned, my father was a gastroenterologist, and unfortunately, he is no longer with us. We often discussed the problems in the gastroenterology space. I come from a technology background, and so does my cofounder, Deepanshu. At the time, I was exploring AI, and we noticed several startups focusing on imaging, like chest X-rays and pathology-based startups. We thought the next natural progression would be to address problems in endoscopy.

I enrolled in an accelerator called Entrepreneur First, where I met Deepanshu. Interestingly, we both are originally from the same place but were settled in different parts of the world—he was in the US, and I was in Hong Kong. I was looking for someone technical who could handle these challenges, and I jokingly asked him, "If you can drive autonomous cars on the roads, can you drive endoscopes into patients?" That's how we started.

We spoke to hundreds of key opinion leaders, both senior and junior doctors. Initially, we started with the key opinion leaders, but the narrative changed when we spoke to trainees and junior doctors. We realized that instead of focusing on aiding experienced doctors, we should address the struggles of junior doctors.

This led us to pivot toward improving the quality of endoscopic procedures. We aimed to question and ensure the quality of the work done by doctors rather than just aiding them. We identified that our target audience would be junior doctors. We avoided common problems like polyp detection, which already had many solutions. Had we followed that path, we might not have survived. Destiny had other plans for us.

Praveen: What does the product do? If it's not polyp detection, then where is the self-driving happening inside the gut?

Saurabh: Self-driving is happening inside the upper GI tract. Specifically, our product focuses on helping doctors with quality control by reducing blind spots during endoscopies. We're not detecting polyps or early cancers directly. Instead, we provide information on the total area covered during the endoscopy.

There are blind spots that are not in the field of view in self-driving cars. Similarly, when doctors drive the endoscope inside a patient's body, they encounter numerous blind spots. Studies show that about 22.5 percent of areas can be missed, leading to suboptimal diagnoses and missed cancers. In the upper GI tract, this is particularly critical because it includes a variety of diseases, not just cancers like in the lower GI.

In summary, our software-based solution helps doctors reduce these blind spots. We're classified as a Class 2a medical device, assisting in ensuring thorough examinations and better diagnostic outcomes.

Praveen: I recall that there's also hardware that goes along. And that's a Graphics Processing Unit (GPU).

Saurabh: Correct. Going back to the self-driving analogy, we use the same GPUs found in self-driving cars. Initially, we were inspired by gaming GPUs because they provided the necessary power. For our purposes, we didn't need the tiny, compact GPUs used in cars; we could use larger machines that offered more computing power.

This approach differs slightly from polyp detection algorithms, as we need several models to run simultaneously to make the final inference. Currently, our strategy involves having a holistic solution with multiple integrated components. This is becoming standard in the field, and many top companies are moving in this direction.

We're focusing on GPU-based solutions due to their current reliability and performance. While cloud-based solutions have potential, they also come with their own set of challenges. For the foreseeable future, we'll be using hardware-based solutions with GPUs and SAMDs (software as a medical device) running on top.

Praveen: How does your product differentiate itself from the other AI companies?

Saurabh: That's a great question. Many companies focus on polyp detection, and while there's potential for overlap, we see a more collaborative approach as beneficial. We focus on what users really want and work backward from there. For example, if a company already has an algorithm for one aspect, we focus on providing something they don't have, like additional quality control or characterization solutions. Our goal is to be part of an ecosystem that offers a comprehensive platform to users. We believe that by creating value together, we can all benefit and make the pie bigger rather than trying to have it all ourselves.

Praveen: Saurabh, one more thing that really intrigued me about your story is the geography and the locations where you're based. This accelerator program where you met your cofounder, where was that?

Saurabh: Yes, the accelerator, Entrepreneur First, was in Hong Kong for about a year. They attract talent from all over the world. My cofounder, who was in Silicon Valley, flew in just for the program. He's now based in India. Being in Hong Kong helped us connect with top clinicians and key opinion leaders. Though we're based in Hong Kong, we focus on Western markets for R&D and eventual launches in Europe and the US.

Praveen: Where are the key opinion leaders you spoke to based?

Saurabh: I started with conferences in India and Hong Kong and then expanded to the US and Japan, especially for endoscopy. We've engaged with clinicians from India, the US, Europe, and Japan. Each region has its unique challenges, giving us a comprehensive view of the problems in different parts of the world and how to address them with technology.

Praveen: Very interesting, Saurabh. Just to recap, you're based in Hong Kong; the company started there, and your cofounder is originally from the US but is now based in India. You have a development team in India, you're rolling out in Hong Kong, and planning to sell in the US and Europe soon. This challenges the idea that healthcare is local. Your story shows that innovation is global. Is there anything else you wanted to share today?

Saurabh: First of all, thanks for your vote of confidence. Quite interesting when I think about it, when you say it now, it's right in my face. That's it from me.

Praveen: I want to summarize the gist of what I'm taking away. There's a quote in AI that's been done to death, which is that it's not AI that's going to take away your job; it's people using AI who would take your job. What I'm taking away from your company is this: it's obvious to me that if a junior endoscopist were to use your product, the differentiation between a junior endoscopist and a well-experienced endoscopist would disappear because he would catch the same things and more with the aid of this tool. People who are going to be left behind are people who are not embracing change and are stuck with their heads in the sand. That's the biggest risk. Congratulations on the progress, and I wish you all the best.

DON'T FORGET THAT THE ROBOT IS PART OF THE TEAM: DR. SANKET CHAUHAN OF SURGICAL AUTOMATIONS

Dr. Sanket Chauhan is the founder and CEO of Surgical Automations. He's a physician innovator with a background in business, medical devices, surgical robots, and clinical simulation. Surgical Automations is developing a robotic platform to help eliminate medical errors and enhance patient safety.

Themes: *robotic colonoscopy, endoluminal navigation, early cancer detection, cognitive and psychomotor skills, regulatory approvals*

Date: *December 2023*

Three Takeaways:

1. **Innovative AI and Robotics Integration in GI Procedures**: Dr. Chauhan's company, Surgical Automations, is pioneering the integration of AI and robotics in gastroenterology, aiming to assist gastroenterologists by automating the most cumbersome steps of endoluminal navigation for procedures like colonoscopy.
2. **The Future Vision for GI Robotics**: The vision for Surgical Automations includes not only improving current GI procedures but also addressing the backlog of colonoscopies and ensuring early cancer detection.
3. **Regulatory and Development Milestones**: Dr. Sanket Chauhan highlights the extensive process involved in bringing medical technology to market, from rigorous testing and data collection to obtaining regulatory approvals from bodies like the FDA. The next significant milestone for Surgical Automations is to conduct

first-in-human trials, with a vision of full commercialization within five years.

Praveen: I've been really looking forward to this conversation and understanding where the future of gastroenterology could go in the context of robotics. What in your background connected you to robotics, and what brought you here?

Dr. Sanket Chauhan: Yes, when I came to the US, I was in a robotics program for my fellowship in Florida at Celebration Hospital, now called Advent Health. I was privileged to be trained under Dr. Vipul Patel. It's been an honor to be trained under him. He's the most experienced robotic surgeon in the world, the founder of the Society of Robotic Surgery, and just a great mentor and friend. That's where my journey in robotics started.

At that time, I always had this mechanical engineering side to me. We were working with the University of Central Florida in the Department of Computer Engineering, collaborating with students to develop new technologies. We actually started a course just for medical robotics. Then, we got a congressional grant from the Department of Defense around 2008-2009, which was focused on surgical robotic telepresence and surgical simulations. This was always what I wanted to do, so I extended my fellowship into that track and became a Department of Defense fellow for a year.

After that, I moved to the University of Minnesota and did another fellowship in surgical simulations. Since graduating, I've been here. We started this company three or four years ago, just around the time COVID hit, and that's where our journey began.

Praveen: What led you to start the company? How did you pick gastroenterology as a focus area?

Dr. Chauhan: We initially did not pick gastroenterology as a focus area. The macro-environment was very supportive of the respiratory field during the peak of COVID. Our core technology is what we call a closed-loop endoluminal navigation system, which is an AI-based system. This system can automate any camera navigation inside a tube within the body, whether it's upper GI, lower GI, cystoscopy, bronchoscopy, ureteroscopy, or renoscopy. Essentially, if there's a tube in the body and a scope goes in it, we can automate that.

At that time, with the macroeconomic environment favoring respiratory applications, we started with endotracheal intubation. It was a perfect fit because everyone understood the importance of ventilation during COVID. Our goal was to demonstrate this technology to VCs and investors, showing what we had achieved and focusing on the larger market potential. We built an automated robotic intubation system, which has great potential, especially in pre-hospital care where EMTs are involved.

Since then, we've shifted our focus more toward gastroenterology, specifically colonoscopy and upper GI procedures. These procedures are very similar mechanically and electromechanically. The market for colonoscopies alone is significant, with about 15 million performed annually in the United States. It's a crucial screening procedure with a huge backlog, making it an excellent area to start with our technology. That's how we transitioned to focusing on gastroenterology.

Praveen: Can you explain the core technology as if you were explaining it to a ten-year-old? What does it do?

Dr. Chauhan: Sure, let's start by understanding how doctors do procedures. We have skills that we can break down into two main parts: cognitive skills and psychomotor skills.

Cognitive skills involve using our brains to identify things inside the body. For example, when we perform a GI procedure, we need to recognize different parts of the digestive system. We learn to identify what a healthy tissue looks like, what a polyp (which could be a growth) looks like, and how to spot things like bleeding. We get good at this because we've seen thousands of images during our training.

Psychomotor skills are about how we move our hands to perform the procedure. It's like when you play a video game, your hands follow what your brain tells them to do. You learn to make precise movements, like cutting or suturing, based on what you see and what you've been trained to do.

Our technology combines these two skills using artificial intelligence (AI). The AI helps with the cognitive part by recognizing different parts of the anatomy and identifying areas of interest, like potential polyps, during a procedure. There are already AI systems in healthcare that help detect shadows in lungs or spots on mammograms, indicating possible issues. Similarly, some systems help identify polyps during colonoscopies.

We take this a step further by using robotics to handle the psychomotor part. Once the AI identifies an area of interest, it sends that information to the robot, which then helps navigate and perform the procedure. This combination of AI and robotics is like having a super-smart assistant that can both see and act, helping doctors perform procedures more accurately and efficiently. This way, we can find and treat conditions like cancer earlier, making procedures safer and more accessible for more people.

Praveen: How does the robot find its way inside the gut?

Dr. Chauhan: That's where our proprietary technology comes in. It requires a lot of data and extensive training. Our team has put in countless sleepless nights to develop this process. The robot learns the anatomy by moving toward specific landmarks, and as it progresses, it encounters new anatomical features, continually updating its understanding. It's a closed-loop system, constantly learning and adapting.

Praveen: Would the data you've trained the robot be anatomical data? For example, for lower GI procedures, would you train the robot on visuals, or would it involve physical simulations?

Dr. Chauhan: It involves different kinds of images, with videos playing a significant role. As we advance, we'll incorporate cadaveric data for additional training. However, we don't see much difference between cadaveric data and the data from a regular colonoscopy performed on a patient. Both provide valuable insights for training the robot.

Praveen: Let's talk about colonoscopy itself. What does the technology do right now, and what's it going to do a year from now and maybe three or five years from now?

Dr. Chauhan: Specifically for Surgical Automations, we're still in the development phase of our technology. To bring it to market, we have to go through regulatory approvals like the FDA in the US or the CE mark in Europe, along with other regulatory environments in different countries. This year, we're aiming to complete cadaver studies, though there's a lot more work to be done before that happens. Medical technology differs significantly from software; you can't release a beta version to get feedback and make quick changes. It has to be thoroughly tested and proven safe before it even touches the first patient. We ask ourselves if the robot is ready to operate on one of our family members, and we know it's safe if we can confidently say yes.

A year from now, we hope to be very close to completing our First in Human trials. This means an aggressive development process will continue. After gathering initial human data, we will move toward FDA submissions and CE submissions. In three years, we aim to be in the process of applying for FDA approval, and by five years, we plan to be fully commercialized.

Praveen: When this is fully commercialized, what is your vision? Let's say this is wildly successful, as you envision. What does that look like?

Dr. Chauhan: Our vision is centered around the patient. We aim for our technology to reduce complication rates, cause less pain to patients, catch cancer early, and reduce the backlog of colonoscopies. Achieving these goals is easier said than done, of course. We have a mountain to climb, and while we have a clear vision of the summit, our focus is on the next step and then the next. If we keep taking the right steps, we'll eventually reach the top.

Praveen: Excellent. And what is that next step right now for the business?

Dr. Chauhan: The next step is to test our technology on cadavers and gather some data. Our team, though lean, is working tirelessly to reach this milestone. We hope to have significant updates by the end of this year and will proceed from there.

Praveen: Excellent, Sanket. Can you give a business update or share any numbers? How much money have you raised? What are you seeking to raise? Any team updates?

Dr. Chauhan: Absolutely. We've completed a pre-seed round, raising close to a million dollars through a note round. Currently, we're finishing up our seed round, which is led by Dr. Fred Moll, a pioneer in the robotic world. He founded Intuitive Surgical, which has a market cap of around $120 billion. He also started Auris, a robotic bronchoscopy company acquired by J&J, and a robotic hair transplant company. Dr. Moll's backing is an honor, providing us with invaluable guidance and experience. After closing the seed round, we'll plan for a Series A next year as we get closer to testing on cadavers and preparing for first-in-human trials.

Praveen: Now, one of the fears that many gastroenterologists have with AI and automation is, "Am I going to lose my job?" Often, people on the AI side are reassured by the quote that "the person using AI will replace those who don't use it, not the AI itself." Let's get past that and delve into the reality. If technology like this becomes available, could a technician, not a clinician or surgeon, perform a colonoscopy?

Dr. Chauhan: Physicians bring 15 years of training and cognitive skills. Being a physician is not just about diagnosing and treating disease – it's also about empathy, the "care" part of healthcare, and seeing patients as human beings, not just as cases. Our physicians are heroes who are unfortunately overloaded with increasing administrative burdens. If we can make their lives even a tiny bit better in the operating room or the endoscopy suite by taking the stress out of procedures and making them simpler, it will have a huge impact on patient care.

Reassuring patients and communicating effectively, especially when delivering bad news, requires a human touch. AI lacks these affective skills. Would you want a robot to tell you that you have cancer? Probably not. The human side of medicine is what makes our work rewarding, and I hope that remains unchanged. How gastroenterologists choose to use robots and whether they involve mid-level providers in procedures is best left to them.

Praveen: If we draw a bell curve, what you describe is ideal. That's how you'd want every clinician to be. But the reality is that not every clinician is on the right side of the bell curve. You're talking about communicating with patients, but most physicians don't have time to communicate (referring to routine communication during consults). Blame the EHRs, administrative burdens, RVU targets, and other distractions. They don't have the time they used to have.

Dr. Chauhan: It's interesting that you mentioned the bell curve. When it comes to quality, there are two systems: lean and Six Sigma. The lean system, like the Toyota Production System, focuses on reducing waste. The Six Sigma system aims to improve reliability, efficiency, and quality. That's what we're trying to do with our robot.

The robot goes from one point to another, which is the mathematically calculated shortest path possible. That is the way it's going to go. No matter if you do it, no matter if I do it, or an APP (advanced practice provider) or a gastroenterologist who has been practicing for 30 years, it's going to go the same way. This minimizes variability.

Healthcare quality isn't just about procedures; it's about the entire system and process. Team-based care is here to stay, and the robot is part of that team. Traditionally, we've had the nurse, technician, patient, physician, and administrative staff. Now, we have intelligent robots, too. Do not ignore them. We can use them for the betterment of patients.

Praveen: Wonderful, Sanket. Do you have any final thoughts?

Dr. Chauhan: Thank you so much for having us, and hopefully, we will have some more announcements by the end of next year.

Praveen: Fantastic, Sanket! I'm so excited about the work that you're doing, and I wish you and your team all the best. I thoroughly enjoyed our conversation.

Dr. Chauhan: Absolutely. And I have copies of *Scope Forward*. Any new hire we have has to read that. The first thing that's a part of their colonoscopy package that we give them is that they have to read this book. Again, thank you so much for bringing these books to life.

DRONES THAT SWIM IN YOUR GUT: TORREY SMITH OF ENDIATX (PILLBOT)

Torrey Smith is the founder and CEO of Endiatx, maker of PillBot™. His background is in aerospace engineering and MedTech. Endiatx is revolutionizing GI diagnosis and treatment with its micro-robotic pill, PillBot™, designed to navigate the human stomach for unprecedented access and care.

Themes: *cross-disciplinary innovation, AI and robotics, MedTech innovation, micro-robotics, molecular machines, drones, early diagnosis*

Date: *March 2024*

Three Takeaways:

1. **Innovative Cross-Disciplinary Approach**: Torrey Smith's transition from aerospace engineering to medical technology highlights the power of cross-disciplinary innovation. By applying principles from aerospace to MedTech, Endiatx has developed a novel approach to navigating the human body with the PillBot, showing how diverse expertise can lead to groundbreaking medical solutions.
2. **AI and Robotics in Healthcare**: The integration of AI and robotics in healthcare, as exemplified by the PillBot, is set to revolutionize medical diagnostics and treatment. AI-enhanced devices can significantly improve accessibility, efficiency, and accuracy in medical procedures, making advanced healthcare available to a broader population, even in remote areas.
3. **Future of MedTech**: The future of MedTech lies in micro-robotics and AI. Endiatx's vision extends beyond

the GI tract to include brain surgery and potentially molecular-sized machines, merging biology with technology. This innovative trajectory points to a future where early diagnosis, minimally invasive procedures, and enhanced patient care become the norm, driven by advanced technological solutions.

Praveen: Torrey, the story of Endiatx and the creation of the PillBot is almost something out of a sci-fi movie. You started in aerospace engineering, and now you're flying something inside the body. I'm really curious, so please tell us what led to Endiatx.

Torrey Smith: Thank you. Like many of us, my inspiration came from science fiction books and movies. Two pivotal movies for me were "The Right Stuff," about the Mercury astronauts, which cemented my desire to become an aerospace engineer, and "Inner Space," which opened my imagination to the idea of tiny devices navigating inside the body. As a child, seeing that made it seem normal. As an adult leaving the aerospace program at Cal Poly, I had an opportunity to go into medical devices. At the time, I was more excited about experimental aircraft, but then my aunt was diagnosed with a glioblastoma, a brain tumor, and sadly passed away. This loss fired me up and made me want to see what I could do by combining aerospace engineering with MedTech.

After working my way through a series of startups in the Silicon Valley Bay area focusing on medical devices, I eventually had the chance to build a team and cofound Endiatx. Our mission is to inject deep tech into MedTech, and we believe that the GI tract and the human stomach might be the perfect starting point.

Praveen: How did you pick GI?

Torrey: Well, let's be honest. Holding the PillBot, an advanced prototype with features I love but might not make it to the first clinical product, is amazing. It has four motors and pump jets, similar to a quadcopter. The one we're entering clinical trials with has three pump jets due to some trade-offs. When I tell people I build tiny robots to put inside the human body, reactions vary from amazement to skepticism.

The human stomach offers a perfect playground. If you drink a large glass of water, your stomach becomes a temporary bag of water, allowing our small swimming drones to navigate and perform tasks. We see the stomach as a starting point for tiny robots in the human body, with plans to explore the rest of the GI tract, blood vessels, and even layers of tissue. Personally, I'd love to reach the brain to address conditions like glioblastomas, completing a full circle for me. The stomach is an ideal beginning, offering untapped potential for accessible screenings.

Praveen: Very interesting, Torrey. What was the initial challenge that you were tackling when you got started with Endiatx?

Torrey: Each founding story is a mix of market research, speaking with doctors, and your own instincts. But there's also an emotional and personal side. After working on several medical device companies, I felt a need for more significant innovation rather than incremental improvements.

When we started Endiatx, we didn't know exactly what our robots would look like. We had many ideas—tank treads, inchworms, flagellating tails, contra-rotating corkscrews, and swimming robots. We realized that navigating the entire GI tract with one device would be challenging. One of our advisors, an Air Force doctor, suggested that if you drank a lot of water, you might be able to swim in the stomach.

We then shifted from the emotional side to market research and found a huge unmet need for accessible upper endoscopy. Combining the emotional drive with market research, we launched Endiatx. My advice to founders is to balance raw emotions with market research.

Praveen: Please describe what the PillBot does today.

Torrey: We're creating a swimming robot drone that's controllable in three dimensions within a stomach. The patient drinks a large glass of water, and the drone uses three pump jets to maneuver. It provides real-time live video to the doctor, who can control it via a touch screen on their phone or a game controller. Our goal is to offer gastroenterologists a highly maneuverable tool for gastroscopy that can be used in a remote setting or a low-barrier entry setting, making quick stomach examinations more accessible.

Praveen: What are the current challenges you're facing?

Torrey: We're about to turn five years old and have worked hard to overcome technical hurdles. We've largely addressed those and are entering clinical trials soon. Now, our focus is on regulatory approvals and market adoption.

Praveen: I want to delve into the technical challenges. I'm trying to visualize a typical gastroenterologist navigating this. Using a gaming device requires a certain tactile skill. Not everyone is used to it. I see that as one potential challenge. Another is how the drone navigates the gut, which isn't straightforward. There's a lot of maneuvering involved. Given what I know about GI patients and emergencies, the gut isn't always clear. What are your reflections on this?

Torrey: How are we going to navigate the small bowel and gut with all its twists and turns? The simple answer is we're not. Our job as a stomach robot is done once we enter the gut. We focus on endoscopy procedures where we see value, like upper GI. Pill cameras work great for small bowel procedures. For upper GI, we see potential. PillBot won't replace an endoscope in the esophagus but can provide value in the stomach, where we see immediate diagnostic relief for conditions like gastritis.

PillBot is inexpensive, and our goal is to show it as a safe and effective endoscope replacement post-FDA trials. With 70 million upper endoscopies annually, we believe PillBot can replace some EGDs and qualify patients for earlier or follow-up EGDs. This could reduce hospital visits and turn many into digital health solutions.

Navigating the PillBot is intuitive. At a recent AGA Shark Tank event, GIs used an Xbox controller to control PillBot within seconds. Feedback was positive, with calls for more resolution and battery life. We aim to make the interface video game-like, with data analytics and situational awareness, potentially integrating tools like an Apple Vision Pro for an advanced user experience.

Our goal is to gather data easily with robot pills and improve control tools, ensuring success by listening to feedback from these amazing doctors.

Praveen: Awesome, what happens after FDA approval and leading up to commercialization?

Torrey Smith: After FDA approval, if we complete trials in 2024 and finish all paperwork in 2025, we could enter the market by early 2026. That's for PillBot Gen 1 in the stomach. But our ambitions don't stop there. We're already planning PillBot Gen 2, aiming for lab-on-chip technology to take samples without removing the robot from the patient. We're also exploring microsurgery capabilities, like polypectomy.

Beyond the stomach, we're looking at colonoscopy, which requires more preparation but offers significant advantages, such as no sedation. Our long-term goal is to pioneer micro-robotics in the human body, aiming for applications beyond the GI tract, including brain surgery.

This journey is driven by our passion for transforming medical procedures and saving lives. We're excited about the potential to offer early diagnosis and treatment for conditions like stomach cancer. Our vision extends to brain-computer interfaces and molecular-sized machines, merging biology and technology.

Ultimately, our goal is to create a new market category and inspire innovation in micro-robotics in medicine. We believe PillBot can make a significant impact on patient care and look forward to contributing to this exciting future.

Praveen: What an exciting adventure! Torrey, I have a couple of questions as we wind up our conversation. What is the funding situation for Endiatx?

Torrey: I'm proud to share that Endiatx has raised over $6 million in our five years as a startup in the Bay Area. With this, we've achieved significant milestones, maintaining capital efficiency.

We're based in a 9,000-square-foot facility equipped with top-tier manufacturing tools, including 3D printers, electronic equipment, mills, lathes, welders, plasma cutters, and even a silent NASA-style air compressor. Recently, we acquired an Arburg injection molding machine, enabling us to handle our own injection molding. We're highly vertically integrated at Endiatx.

Now, as we prepare for clinical trials, we're transitioning from the seed stage to Series A. We're seeking institutional support, strategic partnerships, and family office investments to make a broader impact with our affordable technology. For instance, a PillBot costs about $35 in parts, with an additional $15 for the dongle.

It's an exciting time as we step into the next phase. We've made substantial progress, evolving from large football-sized prototypes to fingertip-sized robots capable of elegant 3D motion. Funding has always been a step-by-step journey aligned with our achievements and investor conversations.

Praveen: Torrey, my customary final question is your reflection on the future of gastroenterology. Independent of Endiatx, where do you see the world go?

Torrey: We're at the tipping point, much like a hockey stick curve where growth is about to explode. This is true in both tech and MedTech. Take Artificial Intelligence (AI), for example. AI can make PillBot 100 times more accessible and affordable. Imagine a nurse in sub-Saharan Africa controlling a PillBot in someone's stomach. AI combines global medical knowledge to create a version of the world's best doctor, extending the reach of human expertise.

I see AI enhancing, not replacing, doctors. For instance, my Tesla can safely drive even when I'm tired—AI should do the same for tedious medical tasks. Doctors, like scientists, should focus on expanding knowledge while AI applies it effectively.

In a few years, AI-enabled Pillbots might be available over the counter, offering diagnostic results and catching early stomach cancers. When something unusual is found, it can prompt immediate medical attention, getting patients to real doctors much earlier than usual.

In summary, AI is crucial for the future of MedTech. I'm excited about the potential of micromechanical systems in the human body, with the stomach being just the beginning.

SECTION: VOICES III - DATA, APPS, AND GAMIFICATION

This section focuses on the transformative power of data, apps, and other innovative approaches, emphasizing the importance of data-driven decision-making in healthcare.

TRANSFORMING GI DATA INTO AN ANCILLARY: OMER DROR OF LYNX.MD

Omer Dror is the founder and CEO of Lynx.MD. His background is in engineering, applied mathematics, and computer science. Lynx.MD's cloud-based platform unlocks real-world health data at scale, enabling secure sharing and analysis of unstructured data.

Themes: *data as a new ancillary, predictive data models, multimodal data integration, physician adaptation, leveraging disruption*

Date: *September 2023*

Three Takeaways:

1. **Leveraging Comprehensive Data for Improved Patient Care**: The future of gastroenterology lies in utilizing a variety of new diagnostic and therapeutic tools along with comprehensive patient data. This holistic approach ensures better treatment outcomes by focusing on complete patient well-being rather than just treating the disease.
2. **Staying Ahead with Technological Advancements**: Gastroenterologists must stay updated with the latest technological advancements to provide the best care. Embracing new technologies and continually improving their skills will be crucial in leading the revolution in patient care.
3. **Importance of Data Integration and Information Flow**: Effective integration of data from various sources is essential for improving patient care. Solving the information flow within the healthcare ecosystem allows for better diagnosis, treatment, and overall patient outcomes.

Praveen: Omer, we are here to talk about why data is the new oil. For context, tell us what Lynx.MD does as a platform.

Omer Dror: Lynx.MD is about making healthcare data accessible. We started the company a few years ago. I come from a cybersecurity and data science background, and I spent many years dealing with sensitive information, managing cybersecurity risks, and extracting actionable insights from data. In healthcare, there's now a massive amount of digital data accumulated within provider systems, thanks to industry-wide digitization over the past decade.

However, leveraging this data to build predictive models, develop new therapies, and evaluate treatment efficacy is still challenging due to several limitations. The data is sensitive, mostly unstructured, and siloed across various healthcare entities. High volumes of data need to be aggregated and made accessible to make this data valuable and improve predictive capabilities.

Lynx.MD addresses these challenges by building trusted data environments. These environments allow healthcare providers to share data securely and privately with scientists, researchers, and developers. Instead of losing control over the data, providers can keep it in a secure environment they control, enabling exploration and model building while preserving privacy and security.

Praveen: Thank you. Why did you choose healthcare? And I know you're working in gastroenterology, so why specifically GI?

Omer: First of all, generally, healthcare is a massive problem compared to other industries. In finance, banking, media, telco, and others, data is already being heavily utilized for insights. Healthcare is really far behind because of data silos, sensitivity, and other challenges I've mentioned. On the other hand, healthcare offers the biggest opportunity to make a difference. Healthcare's clear mission resonated with me. The convergence of all this data presents the biggest opportunity for advancement, which excites me.

Gastroenterology, specifically, is very interesting. It impacts a large portion of the population and is an outpatient specialty that uses multimodal data—endoscopes, imaging, digital pathologies, medical records, and reports. It's a tough problem due to the unstructured data and the need to bring it all together to improve care. Additionally, gastroenterology is fragmented, with most care happening in the community through independent physician practices and PE-backed groups.

To create real value using the data, you need high volumes—hundreds of thousands or millions of patients. Solving the problem of silos is essential. Academic medical centers alone don't see enough patients to fully understand disease progression. Traditional studies with hundreds or even thousands of participants aren't enough. Gastroenterology is fascinating because it involves leveraging community healthcare practices with real-world data to gain meaningful insights.

Praveen: The consolidation in gastroenterology and medicine as a whole, led by private equity, is helping gather this data. Why is this volume of data important for extracting value?

Omer: When you look at a physician's practice, an average physician sees 5,000-8,000 patients. Medicine has a very long tail of conditions and diseases. Many people come in for screening colonoscopies, but the interesting parts that differentiate from the norm are where we learn about better diagnostics. These cases are a small part of the total patient population.

Take biologics, for example. It's a hot area with a lot of research and new drugs. To compare how different treatments affect various patient segments, you need large numbers. You start seeing patterns in classical statistics with hundreds of cases. To look at disease progression and changes over time, you need exponentially larger volumes.

The consolidation of practices into large patient populations, millions of patients, opens up opportunities. It allows us to analyze patient flow and improve workflows and operational processes. Clinically, it helps investigate what leads to better outcomes, reduce variability, and ensure all patients receive the best possible care.

Praveen: Can you share examples of questions that we might be able to ask based on these large data set models?

Omer: Sure. I'll start with operational examples because they often have the most immediate financial impact on practices. One significant issue is no-shows. People not showing up for appointments wastes time and money. We conducted a study on no-shows for a given provider, predicting them days, weeks, and even months in advance. This helps improve scheduling and reduces wasted time.

On the clinical side, an example involves biologics. By examining different patient populations by demographics, location, labs, and medical history, practices can segment patients and predict who will respond better to specific therapies. This can improve treatment effectiveness and patient outcomes.

Praveen: Let's break this down a little bit, right? So you get all these large data sets, or you consolidate data sets from multiple groups, providers, and whatnot. In your platform, do you have several AI algorithms that are using that data and learning from that data, and are you building predictive models? Or are these preset predictive models that already exist that people can look up? Why can't you use generative AI tools to provide these answers?

Omer: First of all, the platform uses a federated model, so we're actually not aggregating all of the data into one place. Every healthcare partner in the ecosystem gets their own trusted data environment. These environments are connected, allowing de-identified data to be looked at collectively while tightly managing compliance and privacy issues, keeping practices in control.

One issue with GenAI and tools like OpenAI's ChatGPT is that data has to be sent to online servers to get responses back, which is problematic with sensitive data. Our trusted data environments, which function as an AI platform, allow you to fine-tune and train large language models on sensitive data within the environment itself. You essentially have your own large language model in that space. It's not only Gen AI algorithms but also traditional AI algorithms.

A key aspect of our platform is the onboarding piece, where we've built AI around data ingestion and understanding. This minimizes the overhead on practices for setup, ensuring a short time to get data into the environment, map it, and catalog it, enabling practices to start building cohorts quickly.

Regarding interfaces, once the data is in the platform, there's a dataset catalog. You can specify criteria, such as IBD patients treated with a specific biologic in a certain geography, to build a cohort. You can use a natural language interface to explore this data, or you can drag it into an analytics interface to start plotting and analyzing patient flows, treatment timelines, lab results, and outcomes.

Additionally, there is a data science interface. The platform allows practices, scientists, life science companies, and other partners to dig into the data based on their level of access.

Praveen: AI has been brewing under the surface for the longest time, and then it's just exploded. Can you lay out the landscape of AI right now as it relates to healthcare and gastroenterology?

Omer: Gen AI is exploding. Everyone's talking about Gen AI, and it has great applications in the medical field. Another significant area is computer vision, especially in endoscopy and digital pathology. These technologies are very impactful, saving time and improving accuracy. Then there's traditional machine learning, which deals with labs and structured data, using that data to make predictions on specific patient populations.

The real exponential growth comes from integrating these three aspects: text and language models, medical imaging, and structured data. When you bring different modalities into one place, AI starts to look at the patient holistically, not just as a set of images. That's what doctors do.

About five or six years ago, Google published a paper on radiology comparing an AI model to radiologists. The AI model performed better on those radiology images, which was groundbreaking. But it didn't give the complete picture because the AI only looked at snapshots of the images. Radiologists have a lot more information when they do their work, which wasn't considered in that study.

Integrating multimodal data, especially when it's sensitive and comes from different systems, is a much tougher problem. When we can tie all these different AI fields together to look at patient populations in a unified way, it will take us to a new level of health equity, reducing variability in care and ensuring patients receive the best quality of care possible.

AI will also help create a learning system where doctors can see when they've done well, when they've made mistakes, and how to improve. This is crucial for improving care. You have to know what happened to the patient to get better. AI will help tremendously with that.

Praveen: How far are we from this holy grail? I love the vision you painted—multimodality data coming from different silos. If you look at the landscape now, there are a bunch of startups in the computer vision space in GI. They're taking endoscopy images, highlighting polyps, classifying them, and so on. That's one set. Then, there's you, analyzing both clinical and business data and extracting insights. And I'm sure pharmaceutical companies have their own data tools and are building something similar. When all this converges, when we get whole-person data and have these AIs talking to each other and understanding the patient as a whole, the insights will be very different. AI's insights will be much more comprehensive. My question to you is, how far are we from that vision based on where the industry is right now?

Omer: We're getting closer. The opportunity is enormous. When you look at these different AI advancements popping up, the limiting factor underneath the surface is the data. An AI company working with video endoscopies needs to work really hard to collect those videos, and once they have access, they can build AI models. But marrying it with clinical data is a whole different level of complexity. The same goes for those processing claims, operational workflows, or medical images and pathologies. To bring all this together, we need to solve the data accessibility issue and handle the sensitivity of the data.

It's simpler than solving the infamous problem of interoperability in the US. We're much closer to it than we thought one or two years ago. The pace is accelerating, and we're moving very fast in this direction.

Praveen: We keep hearing the statement, "Data is the new oil." Why is data the new oil? And what does that even mean?

Omer: The statement that people have been using for a long time, I think the analogy is even better for healthcare. One thing with oil is that it's underground, and you need to pull it out. That's what's happening here. The data is underground, and we need to pull it out to make it work. But also think about the industry—I'll talk a bit about the life sciences side. If I have a drug or a device going to market, it takes 1.2

billion dollars to get a new drug to market. The ability to use real-world evidence to support that process of getting the drug approved is a huge investment. So, if you can reduce the investment, reduce the time it takes, and actually improve the statistics because you can look at much larger patient populations, that's a big deal.

Once the drug is approved, it's all about generating evidence. I want to see which patients benefit from this drug. If I can show that for conditions A, B, and C, my drug is more effective than other drugs, then when you're able to analyze larger cohorts of patients, you can get to a place where you're improving care and reducing costs.

Praveen: The big picture is clear. The clinical need is very clear: it will help patient care and provide economic benefits on the side of big pharma and biotechnology companies. Here's a direct question: why should the average gastroenterologist, generating all this data every single day, care? One question that will pop into everybody's head is, "Who's going to pay for it?" With insurance companies busy reimbursing for managed care models, why should they even bother? What is the economic case here for private practice GI?

Omer: There are four ways of making money here. I'll start with the internal ones. First, if I can look at the data, improve my operations, reduce my operational cost, and keep the same quality or even improve it, I'm saving money. Second, if I have a patient population that is large enough, I can start using the data in negotiations with payers. That's a big thing. Today, the payers have all the strength in negotiations because they have all the data, and the practices basically don't have anything. Once practices are able to take all of that data and prove to payers that they're providing top-level quality care, they can actually have the upper hand in contract negotiations.

The third and fourth ways involve working with pharmaceutical and MedTech companies. You need to get to scale to do this, but there are two types of revenue-generating opportunities here. One is based on real-world evidence—just data licensing—creating a new revenue stream based on all of the data that's created. The second is research. If you have the data and Life Sciences can use it to look at the patient population and see that this is quality data, that actually leads to more research dollars coming into the practice. Some places have research units; some don't. Those that do can drive a lot more research if they show that the data is high quality and that they have the patient population to pass feasibility tests for studies. So, taking all four into account, this is a big money opportunity for practices.

Praveen: That's very helpful. What is Lynx.MD's business model?

Omer: We operate on a software-as-a-service model, so anyone using our software pays a licensing fee. Additionally, we have a data licensing model where we work with pharmaceutical companies, life sciences, and AI companies. These corporations analyze real-world evidence using our data. They pay for the software license to use our software and the data license to access and utilize the data. We partner with healthcare provider systems and practices, especially in gastroenterology, which generates revenue for these practices.

Praveen: Is it necessary for a practice to be part of a large platform to use your service, or can anybody use your service?

Omer: Anybody can use our service. The data licensing piece becomes more interesting if you have a large or unique patient population or unique data assets.

Praveen: Very interesting, Omer, because what you're actually talking about is a digital data ancillary. In the past, ASCs provided ancillary revenues, as did Pathology, Anesthesia, and other channels. What this seems to me is the potential for practices to create a data ancillary. This doesn't exist yet in a comprehensive way; it's happening in bits and pieces, but it's an evolving ancillary, if you will.

Omer: I actually love that analogy. It ties back into the idea that data is the new oil. You need to be able to take advantage of it. IT is usually a cost center, with everyone spending a lot of money on it without generating revenue. This is an opportunity for IT to actually start generating revenue for the practice.

Praveen: If a patient realizes they can get the same level of care or diagnosis from a non-clinician source, whether it be AI or technology, it could lead to significant disruption. In the GI field, I've seen patients become providers with the help of technology, so I see all these disruptions on the horizon. When this becomes mainstream, what do you think will happen to clinicians who are resistant to change?

Omer: They're definitely going to be left behind. I've been in the technology space for over two decades, and one thing about technology is that if you're not up to date in six months, you can lose your position in the industry. While medicine may move slower than other fields, you still have to adapt, grow, and lead. If you're not leading, then you're going to be left behind. If people are too slow to react, the first thing that will happen is they won't be able to provide the same level of quality care as everyone else. Then, they won't be able to negotiate the same contracts. After that, they'll start losing patients. If you're leading, then you're setting the pace, both in your region and globally. And that's where you want to be.

Praveen: Omer, I have one final question for you. You've touched upon the future several times. However, in summary, what is the future of Gastroenterology from your lens?

Omer: With all the new technology being introduced into the field, together with the consolidation and all the data, the future of Gastroenterology is about leveraging all of that disruption to provide comprehensive treatment for patients. We'll have better diagnostic tools, new therapeutic tools identified through data analysis, and a holistic approach to patient care informed by AI-powered insights. We need to move beyond just treating the disease and ensure that patients truly get better. To achieve this, we need all the tools available and a solution for the information flow within the ecosystem. There are many new tools out there, and it's crucial for gastroenterologists to stay on their toes, keep learning, and use everything available to improve patient care. Gastroenterologists, in front of patients, are the ones who will lead this revolution. If they don't lead, they'll be left behind.

FROM MACHINE LEARNING AT UBER TO A STOOL APP: ASAF KRAUS OF DIETA HEALTH

Asaf Kraus is the founder and CEO of Dieta Health. His background is in data science, machine learning, and technology consulting. Dieta Health helps people improve their digestive health using data and machine learning.

Themes: *AI stool analysis, IBS, software innovation, importance of clinical validation studies, patient experience drives innovation*

Date: *September 2023*

Three Takeaways:

1. **Personal Experience Driving Innovation:** Asaf Kraus's personal struggle with severe IBS led to the creation of Dieta Health. His firsthand experience of the inadequacies in current diagnostic methods inspired him to develop a more accurate and comprehensive tool using AI for stool analysis.
2. **Technological Advancements in Gastroenterology:** Dieta Health's platform leverages AI to provide detailed, multi-dimensional stool analysis, which is more accurate than traditional patient self-reporting. This technological advancement is crucial for improving the accuracy of diagnoses and treatment plans in gastroenterology.
3. **Future Vision for Personalized Healthcare:** Asaf envisions a future where technology enables personalized, concierge-level healthcare for all patients, not just the wealthy. By analyzing vast amounts of data from millions of patients, AI can identify the most effective treatments

for specific conditions, reducing the time and resources needed to improve patient outcomes.

Praveen: Asaf Kraus, founder and CEO of Dieta Health, a very warm welcome.

Asaf Kraus: Thanks, Praveen. It's great to be with you. I've listened to a lot of your episodes with some of the big innovators in gastroenterology, and I appreciate everything you do for the community. So, I'm excited for this conversation as well.

Praveen: When did you actually found Dieta Health?

Asaf: I started thinking about it while I was still at Uber in 2017 and 2018. We really got started with the company in early 2019. So, it's been about four and a half years, and it's been quite an amazing journey. I've learned a lot along the way.

Praveen: You were an engineer at Uber doing machine learning. Who would think about taking stool pictures and applying machine learning to stool pictures of all things? I know I'm making light of perhaps a personal situation, but I do want to learn more.

Asaf: It's definitely okay to make fun of it. It's something we get often, and it is pretty random to people when they first hear about it. The original story of how we came to do this is very personal to me and came from my own experience with gastroenterology.

In early 2017, I was working as a data scientist at Uber, immersed in an entrepreneurial environment. I loved analyzing billions of Uber trips and using the data to predict things.

Then, I was afflicted with severe IBS. It gradually crept up and then became severe, taking over my life. I started sleeping only four hours a night, spending two to three hours a day on the toilet, making multiple runs but unsuccessfully having bowel movements. It affected my dating life and ruined a vacation to Japan, where I had to come back early for a colonoscopy. For about ten months, IBS completely stole my life.

During this period, I was constantly consulting gastroenterologists, my primary care doctor, and dietitians. I ordered dozens of supplements from Amazon, tried all the prescription medications, and experimented with radical diets. My life became focused on solving my IBS and maintaining my job at Uber for health insurance. That's where the idea for Dieta Health began.

One pivotal moment was when my gastroenterologist told me to record a journal of my eating habits and bowel movements, pointing to the Bristol Stool Scale. This one-through-seven diagram categorizes stool from pure constipation to diarrhea, with four being the perfect bowel movement. My stool didn't fit neatly into the scale. There are multiple variables—how liquid or solid the stool is, whether it's fragmented or consolidated, the size, the clarity, and the frequency of bowel movements. Diarrhea and constipation are one-dimensional descriptors, but stool has multiple dimensions.

Later that year, I participated in a clinical trial for a pharmaceutical company. I noticed that patients were asked to look at their stools and record them on a physical survey, which the FDA would use to evaluate the drug's effectiveness. As a data scientist, I realized that computers using computer vision could analyze stool images more accurately and objectively than humans. This realization formed the backbone of Dieta Health and our stool image recognition AI.

Praveen: Very fascinating and inspiring. I see this trend across the board now, not just in gastroenterology, where patients are becoming providers. You're not a clinician, but you are providing care by putting together a system that doesn't exist in our existing system. You put together a program, a system that solves the problem for other patients like you. Now, do you observe this trend as well, and do you have any reflections on that?

Asaf: Definitely. We see a lot of patients becoming entrepreneurs in the community. Companies like Google, Facebook, Meta, Uber, and other Silicon Valley firms have many employees dealing with health issues. One out of five Americans has IBS, and one to two percent have IBD. Within these companies, many people see machine learning and other relevant technologies around them and get inspired to apply them in the medical space.

This trend is very positive. There's more demand for gastroenterologists than there are available specialists. Seventy percent of IBS patients live with it undiagnosed and don't seek care, often due to shame and embarrassment. They'd rather search online in incognito mode for solutions than admit their problems to another person. Technology can step in and help address these issues.

Praveen: Talk a little bit more about how Dieta Health itself works as a platform. The platform is not just stool image recognition.

Asaf: Sure, the stool image recognition is a core part of it, but the app guides you with a cartoon toilet seat on how to photograph your stool. As a patient, you take a picture, which is anonymous unless you choose to provide your identity. It's encrypted, secure, private, and not saved on your camera roll, so your friends won't accidentally find pictures of your stool. This data is stored in the backend and can be transmitted to your gastroenterologist or care team.

The AI we've built classifies your stool based on five objective clinical data points validated in clinical trials, including the Bristol Stool Scale and four other granular characteristics like liquid to solid, presence of blood or mucus, stool clarity (relevant for colonoscopy prep), and stool color (relevant for liver and pancreas diseases). We're developing more characteristics continuously.

Beyond stool image recognition, we've built a comprehensive platform for managing telemedicine interactions in gastroenterology care. The app includes a chat component where patients can communicate with clinicians, and they can also take photos of their meals. Our computer vision categorizes these meals, allowing patients to see the relationship between their diet and stool.

The app also has a survey mechanism, and there's a clinician portal. Gastroenterologists and researchers can use this portal to view patient data, analyze it, and export it for various purposes.

Praveen: How does the company or what you do differ from other startups that are approaching IBS?

Asaf: As far as we know, we're the first and only clinically validated stool image recognition AI available on a mobile app. There are other companies that make smart toilets with the hardware you install, and there are pros and cons to each approach. The advantage of a smart toilet is that the angle, lighting, and conditions for taking pictures are more standardized, leading to uniform data.

However, the biggest drawback of smart toilets is the cost and the fact that they require installation in one place. Our solution, being on a smartphone, is much more accessible. You can download it easily with a QR code, making it very low-cost and convenient. For example, if a pharma company is running a clinical trial and needs two months of stool image results from patients taking an experimental medication, they can distribute the software to the patients' phones easily.

Additionally, if a patient needs to use it before a colonoscopy to ensure their prep is successful, they only need the software on their phone for a few days before the procedure. This makes our approach much more flexible and scalable compared to hardware-based solutions.

Praveen: What's your business model?

Asaf: We're targeting two main business segments: pharmaceutical companies and academic medical centers running clinical trials and providers performing colonoscopies. We started with the pharmaceutical and clinical trial space, partnering with Cedars-Sinai Medical Center through their accelerator program. Our first clinical trial there demonstrated to a pharma company that our app provides more accurate stool data than patient self-reporting. This is crucial, as over a thousand GI clinical trials each year rely on patients' observations of their stool.

We compared our AI to patient observations and two gastroenterologists, showing that the AI is more accurate. Our app helps pharma companies get detailed, accurate data, improving their chances of bringing drugs to market by understanding patient subgroups better.

Recently, Dr. Manoj Mehta in Chicago suggested using our stool image recognition for colonoscopy prep. We've started clinical trials and plan to launch this product line in early 2024. With 200 million colonoscopies worldwide annually and a 10 to 25 percent improper prep rate, there's a huge need. Our app guides patients through the prep process, provides 24/7 support, and reduces the burden on gastroenterologists. We're packaging this as software and service, selling it to providers performing colonoscopies.

Praveen: What did you learn from the clinical studies you conducted?

Asaf: The first major validation study with Dr. Mark Pimentel at Cedars-Sinai was pivotal. When I introduced the idea of photographing stool, he said, "I don't trust my patients." He's been running GI clinical trials for 20 years and has to rely on patient self-reporting, which is subjective. He and Dr. Ali Rezaie annotated 250 stool images on five dimensions, providing the source of truth. We found that our AI was far more accurate than patients at predicting what a gastroenterologist would.

What surprised me most was the patients' willingness to photograph their stools. I took photos of my own stool for two years to build the technology, so I knew I was committed. But seeing other patients do it when instructed after watching a short animated video was pleasantly surprising. This behavior showed the feasibility of our approach.

We also ran clinical trials with Dr. Gil Melmed on ulcerative colitis patients at Cedars-Sinai. We found significant correlations between Dieta's stool analysis and blood CRP levels, suggesting non-invasive monitoring of IBD disease activity.

Additionally, we conducted validation studies on cirrhosis patients at the VA in Virginia and the Mayo Clinic with Dr. Jasmohan Bajaj and Dr. Douglas Simonetto. These patients used our app to guide their lactulose dosage, and the AI proved more accurate than patient self-reporting. The study was presented at the Liver Meeting in Boston.

We're expanding into other areas, like oncology, for opioid-induced constipation. These studies show that our AI can effectively measure stool outcomes across various diseases, providing valuable, accurate data.

Praveen: Fascinating. I can easily see it spreading in many different directions because stool would be relevant to many disease conditions. Do you have any numbers to share, such as how many images or patients?

Asaf: We currently have about 300,000 annotated stool images from around 25,000 patients. The data set includes a variety of patients—ulcerative colitis, IBS, constipation, cirrhosis, and even healthy individuals. The app has progressively improved over time, enhancing user experience, classification accuracy, and variable tracking.

We're also tracking how stool changes when medication is introduced and observing fluctuations in stool characteristics. We're excited about expanding into more medical centers and starting new clinical trials. For example, one upcoming decentralized trial will include 10,000 patients.

Praveen: When you're identifying or categorizing something, you might start with five types, like the stool scale, but AI can find finer gradations that the human eye might miss. Have you had any insights from this? When you scale to millions of data points, could we even evolve common clinical measures?

Asaf: There are two ways to answer this: granularity of each variable and clustering of many variables. For granularity, we developed the AI with the idea that it could be more detailed than a human. I started the annotations myself using an annotation portal. For example, we look at stool consistency on a scale of 0 to 100, where 0 is liquid and 100 is solid. We initially annotated in increments of ten because it's easier for humans to make those decisions. With enough data, the AI can predict continuous variables, understanding what 57.2 looks like, for instance.

On clustering, we made three clusters of IBD patients for a pharmaceutical client using a twelve-dimensional clustering of stool characteristics and their variances. We looked at how fluctuations in stool characteristics, like fragmentation or consistency, could predict who would benefit from a medication. This isn't something the human eye can detect, but with tons of data, a computer can identify clusters of patients who respond well to certain treatments. This has a lot of promise for personalizing digestive health, which is our mission.

Praveen: Asaf, I have a couple of questions about your technology vision. With all these advancing exponential technologies, like augmented reality, one thought that occurred to me is when you point your smartphone to the stool and take pictures, it's currently working in an offline manner. You upload the pictures, and the AI analyzes them later. But as smartphones get augmented reality capabilities, you might be able to take pictures and have the AI analyze them instantly. What are your thoughts on this, and what's your broader technology vision for Dieta Health?

Asaf: Actually, the Dieta app already gives feedback on your stool within ten seconds. The images are uploaded to the cloud and analyzed there, and then the results are sent back to your phone. We're currently working on new technologies that allow real-time analysis. For example, one of our engineers is implementing a new reactive camera framework for mobile development. While our larger AI model that classifies five validated stool characteristics won't fit on a phone due to its size, we've developed simpler models for real-time use.

For instance, we've created a model that distinguishes between stool and non-stool images. This is useful for patients with hepatic encephalopathy, who might accidentally take pictures of the floor. The model can give immediate feedback, prompting the patient to retake the picture if necessary. This kind of real-time feedback is crucial.

Real-time notifications and alerts to clinical teams are another valuable application. If a patient who previously had no blood in their stool suddenly starts showing blood, the app can immediately notify both the patient and their gastroenterologist. Similarly, if a patient's stool turns gray, indicating a potential emergency, the app can alert them to seek immediate medical attention. These alerts can help catch serious conditions early.

Regarding your question about AR (augmented reality) and our larger technology vision, my experience with IBS and the random experiments I went through inspired me. The future of gastroenterology should involve patients knowing more about their own data. Every day, new patients are diagnosed with IBS or discover their IBS. In the future, they should be able to give their data to a system that compares them to millions of other GI patients and identifies the most effective solutions for their specific condition.

We're focusing on stool image recognition to understand sub-diagnoses of IBS better. By analyzing stool images over a few days, the system can compare a patient's data to similar cases in our database and suggest effective treatments based on that comparison. This approach aims to improve patients' quality of life faster and more efficiently. That's the big vision for Dieta.

Praveen: Kudos to you and your team on that vision, and all the best for making that happen. One question that keeps coming up in the private practice GI space is, "Who's going to pay for this?" Given the constraints of managed care and reimbursement models, how did you figure out funding? What advice do you have for those wanting to innovate but facing this question?

Asaf: It's crucial to talk to stakeholders first. I'm still learning this every day. Instead of blindly developing technology, it's better to understand who will pay for it before building it. I often use Figma to design app interfaces and show them to potential stakeholders to understand how it impacts their incentives.

For us, pharma and research have a willingness to spend on capturing superior data because of clear problems, like errors in clinical trials due to patient-reported stool observations. In the colonoscopy space, we identified the high costs of unsuccessful procedures. We are measuring these costs in our clinical trials and aim to publish our findings soon.

We've also been supported by two amazing accelerator programs: UnitedHealthcare Techstars and Cedars-Sinai Accelerator. These programs introduced us to key stakeholders and taught us how to raise money. After participating, we raised a $2 million pre-seed round, which kick-started the company. I highly recommend these programs for anyone looking to innovate in this space.

Praveen: Asaf, as we wrap up our conversation today, my last question to you is, what is the future of gastroenterology?

Asaf: I see a future where it takes fewer resources and less time to move a patient from extreme digestive suffering to thriving in their life again. Currently, if you're wealthy, you can assemble a team of doctors to give you personalized care, read all the clinical trials for you, and curate your treatment. I believe technology can bring this level of care to everyone. Just like Uber democratized the experience of a black car service, technology can make concierge healthcare accessible to the masses. The reduction in resources required to deliver personalized treatment will be driven by advancements in AI.

Praveen: On that note, thank you so much for sharing these insights. It was great to reflect on the future of GI with you.

Asaf: Thanks, Praveen. It was a pleasure to talk with you today and always. I appreciate everything you're doing.

UNLOCKING THE VALUE OF DATA FROM GI SURGERY CENTERS: DUNSTON ALMEIDA OF TRIVALENCE

Dunston Almeida is the founder and CEO of triValence. His background is in healthcare, FinTech, and M&A. triValence is a B2B healthcare payments network for medical device manufacturers, distributors, and providers.

Themes: *data-driven efficiency, private equity consolidation, surgery centers, cost management, procurement automation, surgical supply chain*

Date: *November 2023*

Three Takeaways:

1. **Data-Driven Efficiency in GI Surgery Centers:** TriValence automates the procurement and payment processes, reducing operational inefficiencies and providing real-time data insights for surgery centers. This approach helps surgery centers manage costs, which is crucial in an environment where reimbursement rates are stagnant while operational costs are rising.
2. **Future-Proofing GI Practices through Scale and Innovation:** The GI industry is poised for significant consolidation, driven by private equity and other investors seeking to optimize payer contracts and operational efficiency. GI practices that can leverage data for strategic decision-making and adopt innovative technologies will be better positioned to succeed.
3. **Strategic Use of Data as a Competitive Advantage:** Dunston emphasizes the importance of utilizing the vast amounts of data generated by GI practices to gain a

competitive edge. By analyzing and leveraging this data, practices can enhance patient outcomes, negotiate better contracts, and streamline their operations.

Praveen: Firstly, what does triValence do as a company?

Dunston Almeida: Fundamentally, we said that triValence would try to fix this massive inefficiency across the entire surgical center industry in the way in which they order and pay for all the products and services they need to perform surgeries or procedures. We found that the ASC industry hasn't adopted technology and automation the way hospitals or health systems have. When we looked at the amount of waste and inefficiency in ordering supplies, goods, materials, managing them accurately, billing for them, invoicing them, and then paying for them, we found that the ASC industry, with GI being one of the subsectors, was further behind than we thought. We asked ourselves if there was a much cheaper, faster, better way to do this on behalf of all the providers in this industry. This isn't a payer-provider battle. This is just about getting some operating inefficiencies out of the way and allowing the GI industry, for example, to do more procedures and offer better patient outcomes.

Praveen: Is this a software platform that you have built? How does it do what you said it does?

Dunston: Partly a software platform. We built the only B2B network connecting every major medical device manufacturer, distributor, and the ASC industry. We're electronically connected to companies like J&J, Medtronic, Stryker, Zimmer, McKesson, Cardinal, and Henry Schein. We're also connected to many ASC providers and GI practices. The software automates purchasing everything from bandages to scopes and implants. We then take all that information and electronically pay the suppliers, providing real-time data reconciliation. This means we can show if the invoice, purchase order, and payment match, which is crucial because there's often a discrepancy between what was used, billed, and paid for. Our system captures all this data electronically, providing real-time insights into trends and performance. This data helps surgical centers manage their business better and helps suppliers understand market dynamics and optimize their sales efforts.

We've also built our own payment network to handle electronic payments, as most surgical centers still use checks. This network gives real-time data, making payments more efficient and accurate, reducing order inaccuracies, and improving working capital management for suppliers. Our solution flags potential errors before they happen, ensuring smoother operations. Although we're a young company, we've seen great adoption and feedback, especially from leading players in the GI industry.

Praveen: So, to understand, triValence is like Amazon for surgery centers, connecting buyers and sellers through a network software platform and handling payments to ensure accuracy and prevent errors. Is that right?

Dunston: Exactly. It's a great analogy. B2B purchasing in healthcare is complex. On Amazon, you search for shoes and get thousands of options. In healthcare, we have millions of SKUs. We've aggregated product data down to the SKU level, ensuring you know exactly what you're looking for. B2B complexity includes GPOs, contract prices, and special vendor pricing. We don't compare prices like a marketplace, but we make sure all contracted prices are included. We harmonize data from GPOs, device companies, and distributors. When you, as a physician, want a specific surgical screw, we know exactly what you need. This removes the friction of manual ordering and ensures accuracy and efficiency.

Praveen: You started triValence during COVID. What in your background led you to this, and how did you build the company so quickly?

Dunston: My dad, a small practice orthopedic surgeon, was tech-phobic, and I saw the inefficiencies firsthand. With my experience in payer IT, provider side, and pharmacy distribution, I saw a huge opportunity. During COVID, I wondered why the healthcare supply chain still relied on paper and phone calls for millions in purchases. Hospitals have invested heavily in EMRs, but ASCs, being nimble, needed cheaper, efficient solutions. Building triValence was challenging due to the need to connect diverse data and electronic payments. We spent over a year refining our tech with feedback from clinicians, ensuring it met their needs. Even though healthcare adopts technology slowly, focusing on the customer experience for doctors, providers, and surgical centers is crucial. Our platform addresses the inefficiencies in purchasing, payment, and data across various specialties. GI, for example, has a high volume and needs better cost management. Despite being only two years old, leveraging existing tech like PayPal and Venmo helped us accelerate development. Our expertise tailored these solutions for the surgical center industry.

Praveen: Given your background, how do you see things playing out in GI over the next three to five years? You're representing a financial, private equity, and hedge fund background.

I'm curious to know, from your viewpoint, how you see the business of gastroenterology playing out in the future.

Dunston: First of all, it's a great question, Praveen, and it's fascinating because I'm going to give you an example of radiology. About eight to ten years ago, when I was in the payer IT side of things, I was part of a company that managed a lot of imaging and diagnostic facilities. AI wasn't very cool then, but I acquired a natural language processing company for Mass General that looked at unstructured data in medical records and tried to synthesize it. The goal was to get a true longitudinal picture of a patient and possibly be preventive. What happened in radiology was similar. There was a lot of consolidation and private equity roll-ups, and now, in the past six to twelve months, people are talking about using AI to improve diagnostics and teleradiology. They've faced the same pressures: declining reimbursements and commoditization. They're all trying to figure out how to distinguish themselves.

For GI, the parallel is clear. In five years, you'll see continued adoption by private equity and other third parties wanting to roll up the industry for better payer contracts and improved revenue cycle management. The question becomes: what distinguishes you? What makes your practice stand out in a world where AI is becoming pervasive? The real question for many in the GI landscape is differentiation. Is it through a better patient experience, where patients can get a quick procedure, pay electronically, and have a great experience with their doctor? Or will GI become institutionalized, like radiology, losing that patient touch?

Clearly, GI is not there yet, but consolidation is inevitable. The winners will be those who create a unique, efficient, scalable business model. Who can provide a value-based contract, secure great fee-for-service contracts, and maintain efficient operations? This includes revenue cycle, purchasing, and back-office management, ensuring the system can scale without failing. The term Wall Street uses is "institutionalization." We're seeing the evolution from family practices to large numbers of sites. What is the right size to make a GI company transformative? If you look at SCA or Surgery Partners, they started with orthopedics and expanded to become almost brand names. The real question is: can we create a GI brand name that brings a unique experience?

Praveen: Dunston, as we wrap up our conversation, any parting thoughts?

Dunston: I just want to leave people with the value of the data that GI has. It's one of those rare industries where you have a wealth of information. If you think about what you can do in your practice and business today using the information you have, even if it's on paper, what can you do to give yourself a competitive advantage? In a world facing increasing reimbursement pressure, where Medicare Advantage is under massive CMS pressure, and where reimbursements might go up slightly but your costs are increasing much faster, what can you do to position yourself better? That's the existential question we should all be thinking about in healthcare.

Praveen: I'm so glad we're ending on this note because data is indeed the asset that everyone in healthcare should be focusing on. Thank you; it's been very insightful talking to you.

Dunston: Thank you, Praveen. Great questions!

GAMIFYING GI DATA ANNOTATION FOR AI: ERIK DUHAIME OF CENTAUR LABS

Erik Duhaime is the founder and CEO of Centaur Labs. His background is in collective intelligence, human evolution, and biology. Centaur Labs accelerates AI developments in science and healthcare by providing high-quality, scalable data annotation for healthcare data, leveraging collective intelligence and pay-for-performance incentives.

Themes: *human-AI collaboration, medical data annotation, gamified data opportunities, algorithms, monetizing medical data*

Date: *August 2022*

Three Takeaways:

1. **Human-AI Collaboration for Enhanced Accuracy:** Erik Duhaime's company, Centaur Labs, leverages the concept of human-computer collaboration to enhance the accuracy of medical data annotation. By combining the inputs of semi-experts and AI algorithms, Centaur Labs outperforms traditional methods of data annotation. This approach highlights the potential of integrating diverse human skills with advanced AI to achieve superior outcomes in medical diagnostics.
2. **Scalability and Efficiency in Data Annotation:** Centaur Labs addresses a critical bottleneck in AI development for the medical field: the need for large, accurately annotated datasets. Their gamified platform, DiagnosUs, engages medical students and professionals to annotate data efficiently while maintaining high-quality standards. This innovative solution not only meets the data needs of AI

companies but also offers educational benefits and financial incentives to participants.
3. **Future of Medicine: Networked Collaboration:** The future of medicine, as envisioned by Duhaime, involves a networked approach where people and AI systems collaborate seamlessly. Instead of viewing AI as a replacement for human doctors, the focus shifts to creating integrated workflows where AI aids in diagnostics and human experts provide oversight and additional opinions. This model aims to improve diagnostic accuracy, enhance efficiency, and ensure comprehensive patient care by leveraging the strengths of both human and machine intelligence.

Praveen: Erik, since the first time we interacted and you came as a guest on the GI Mastermind, I've been looking forward to this conversation. Let's start at the beginning. How did you get started?

Erik Duhaime: I was doing my PhD at the Center for Collective Intelligence at MIT. If I go even further back, my undergrad and master's thesis were both on the evolution of cooperation and altruism. I was really interested in when a group of people is greater than the sum of its parts—how to combine people's opinions or coordinate them in new ways to achieve that. At MIT, I got bitten by the tech bug and became interested in not just combining people but also using information technology to combine people's opinions with artificial intelligence algorithms.

I started running experiments to see how to best combine people's opinions with each other and with AI algorithms. During my PhD, I discovered that groups of semi-experts, like med students, could classify skin lesions and, if I overweighted the high performers and combined their opinions in different ways, they could outperform board-certified dermatologists with ten years of experience. Even more exciting, I found that combining these groups of semi-experts with a state-of-the-art AI algorithm could outperform either the algorithm or the dermatologists alone. In fact, these combined groups outperformed board-certified dermatologists paired with the AI algorithm.

This idea of human-computer collaboration is why we're called Centaur. The initial findings showed that when combining multiple inputs—whether people or computers—you want to identify complementarities and how the group should be formed. It's not just about having the five best people; it's about having diverse skills, like putting together an optimal trivia team with different strengths. Because people and computers see problems differently, the best team is often not made up of the best individual performers.

Praveen: That's very interesting. Can you contextualize what the company does for a specialty like gastroenterology?

Erik: Yeah. When I was starting the company, based on my PhD research, my first thought was, great, we're better than dermatologists. Let's cure cancer and be diagnosticians. However, we received feedback about the potential legal issues, reimbursement challenges, and other obstacles. That's when we stumbled into the data labeling market for companies developing AI.

To develop artificial intelligence, you need large datasets of accurately annotated data. I often explain this by bringing up CAPTCHA, where you have to identify crosswalks, trees, and cars to prove you're not a robot. Google uses this to train self-driving cars. In the medical domain, it's similar. If you want an algorithm to spot polyps in a GI video, you need many videos where the polyps have been annotated in each frame. This isn't just ten patients' worth—it's hundreds or thousands of videos. Hiring GI docs to do this work is extremely expensive and tedious.

This challenge is the same across various domains where we work, such as EEG data, robotic surgery videos, chest X-rays, dermatology, and more. One of the main bottlenecks in developing medical AI is accurately annotating these large datasets. That's where we come in. We help companies annotate large datasets by leveraging a network of tens of thousands of medical students, doctors, and other professionals worldwide. They use a gamified competitive app to annotate data, ultimately aiding medical AI companies in developing new technologies.

Praveen: Can you please describe what the Centaur platform does?

Erik: Yes. We have a web portal where customers can upload their data, scope out their projects, see the results come in, and do some QA. However, the core product is a mobile app called DiagnosUs. Anyone can download this on an iOS device. The beauty of it is that it's gamified. It's a competitive, gamified data annotation app where you earn money if you're better than other people.

Starting out, we'll give you some tasks. If you aren't any good, we don't trust you, you don't win any money, and we don't count your opinion. The way it works is we'll launch these competitions, saying, "Hey, you need to look at 100 frames from an endoscopy in the next ten hours to qualify." We mix in gold-standard images created by multiple experts. As a user, you get feedback on every other case, learn if you're doing a good job, and earn points based on your accuracy. If you're the most accurate, you show up as number one on a leaderboard. After completing 100 cases, you see your rank on the leaderboard, and if you're in the top positions, you earn money.

Quality control is built into the system. We don't need another expert to oversee and give feedback. Our users participate for various reasons. Many med students like getting feedback and experience on cases. I started the company when my wife was in med school and saw her paying for flashcard apps, which struck me as crazy given her educational experience could also be valuable. So, many medical students use it to see different images and get feedback.

We also have people who are in it for the money. You can win hundreds or thousands of dollars on the app. It's not a retirement plan, but considering people spend hours on Instagram or TikTok, many residents or fellows use the app during overnight shifts to earn some money. Lastly, some people are in it for the game. I like playing chess and strategy games on my phone, and we have doctors who play because they enjoy seeing themselves on a leaderboard.

Overall, the platform is a gamified, competitive app where the game engages people, making them do many tasks for little money because they get educational and other benefits. Quality control is built into the process because we know how well someone is performing at each task. It doesn't matter if you're a GI doc; what matters is if you're doing a good job at the annotation task right now. The competitive framework ensures only the top performers get paid.

Praveen: A few months ago, I attended a conference at MIT focused on AI and data. The takeaway was that clean data is the hero, not the algorithm, and there's a lack of data. The data world was divided between companies like Centaur, which cleans and labels data, and companies generating synthetic data for AI. Is synthetic data competition for you? Could you explain it a bit more?

(Synthetic data is artificially generated data that mimics real-world data for use in various AI applications.)

Erik: Synthetic data companies use general adversarial networks to create images that machines can't distinguish from real ones. This generates many high-quality, representative images and can address some privacy concerns, especially in the medical domain. However, I'm cautious because it's still early days for this technology.

For example, in dermatology, there are biases between dark and light skin in how algorithms work. Some companies generate synthetic cancerous skin lesion examples on dark skin to create a more balanced data set. However, generating synthetic data might replicate the biases we're trying to eliminate because it's based on limited examples.

Similarly, there's potential in automating and reducing the labeling burden by developing an algorithm that labels most data and allows humans to review uncertain cases. However, the idea that you only need 20 cases to label every image in the world is unrealistic. Models trained on a few examples won't perform well in diverse real-world settings.

I'm skeptical of both pre-labeling and synthetic data as ultimate solutions. Even when algorithms become better than doctors, combining people and computers will still outperform either alone. I don't think AI will replace doctors in most tasks in the midterm. We can't entirely remove the need for human labor, even with good training data sets.

Praveen: From the perspective of your clients, what are they doing with all this annotated data? Whether it's in GI or other specialties, what's their ultimate goal?

Erik: While I can't discuss many of our clients, I can share an example from Eko, a digital stethoscope company. Eko captures heart and lung sounds digitally, allowing them to record and analyze these sounds. They had tens of thousands of recordings and wanted to build an AI algorithm to assist doctors and even less skilled individuals in various settings.

To develop their algorithm, they needed to annotate these recordings. Our network went through these clips and classified them for various heart murmurs, wheezing, coughing, and other sounds. Eko's data scientists used these annotations to train a machine-learning model. This process improved their model's accuracy, and they now have an FDA-approved product on the market, helping clinicians diagnose conditions worldwide.

Praveen: Erik, in these last few years since you've been running the company, what have been your big insights? Did anything surprise you with this whole model?

Erik: One surprising insight is the value of understanding uncertainty and problem difficulty. We get multiple opinions on every label, so we know which cases are confusing. For instance, 90 percent of cases might be straightforward, but for the remaining 10 percent, opinions vary widely. This insight helps clients identify edge cases and refine their models.

For example, a model developer might not consider a scope covered in blood, which could throw off the algorithm. Knowing which cases are tricky allows us to address these issues before scaling. My cofounder, Tom, who worked at Cruise, a self-driving car company, shared a story about late-stage realization that ambulances need different labeling than cars. Early identification of such issues is crucial.

Another example is a company labeling fireplaces for insurance purposes. They had to decide if a TV screen showing a fire counted as a fireplace. Multiple opinions would highlight this discrepancy early on, preventing model errors. These insights into case difficulty and edge cases provide immense value to our clients.

Praveen: How could a private practice gastroenterologist monetize the data they generate every day? Any thoughts or ideas?

Erik: We partner with many data brokers because our clients' developing algorithms need both data and labels. These brokers work with private practices to access and license their data to research companies. This is crucial because algorithms need diverse data sets to be effective. Often, algorithms trained at a single institution like Stanford may not work well elsewhere.

Private practices, though often sitting on valuable data, may see it as a burden due to storage requirements. However, this data is a valuable resource, especially for training robust AI algorithms. If private practices want to contribute to AI research and monetize their data, they should consider partnering with companies that aggregate diverse data sources. There are a few companies working on this, though no clear leader yet. If interested, reach out, and we can connect you with relevant partners.

Praveen: Erik, as my final question, what does the future of medicine look like, especially for gastroenterology?

Erik: I've touched on parts of this, and it relates to our company's central idea and the MD versus AI debate. Many people fear AI, viewing it as a competition between human doctors and algorithms. However, the trend is shifting toward seeing AI as a tool that doctors can use. This comforts some, but issues arise when the doctor and the algorithm disagree. Often, people default to trusting the doctor and ignoring the algorithm, rendering the AI's input pointless. Conversely, if you only trust the algorithm, why involve the doctor at all?

What's missing is the realization that new tech like AI doesn't just replace work; it changes how work is organized. In the future, it won't just be an algorithm or a doctor or even a doctor with an algorithm. We'll see networks of people and computers collaborating to solve diagnostic and data annotation problems. Imagine a colonoscopy video where an algorithm flags areas of interest. Instead of a doctor merely signing off on these flags, controversial or uncertain findings could be quickly shared with other experts globally for additional opinions.

Take the example of the Netflix Prize. Competitors realized they couldn't win individually, so they combined their algorithms and shared the prize. Often, combining multiple algorithms yields better results than a single one. In healthcare, we could see a network where different algorithms and human experts work together to solve complex problems efficiently.

There needs to be a system between an overly sensitive heart detection algorithm on your Apple Watch and a $500 cardiologist visit. These workflows, where AI and human expertise dynamically interact, are the future. Our company is well-positioned to play a role in this shift, but we're not there yet. People are still focused on creating safe algorithms rather than developing these integrated networks.

Praveen: Wonderful, Erik. I found this whole conversation fascinating and insightful. Are there any final comments you want to leave us with?

Erik: I don't think so. I'll just leave you with a thank you and encourage people to reach out. Thank you.

SECTION: VOICES IV - NEW CARE MODELS

This section explores new models of care in gastroenterology, including hybrid and digital health approaches, and their potential to improve patient outcomes.

FROM GI PATIENT TO DIGITAL HEALTH CEO: SAM JACTEL OF AYBLE HEALTH

Sam Jactel is the founder and CEO of Ayble Health. His background is in science, venture investing, and management consulting. Ayble Health combines a multidisciplinary care team, clinical protocols, behavioral science, and machine learning to create personalized care for each patient.

Themes: *personal experience as a patient, ulcerative colitis, patient-driven innovation, precision health, digital therapeutics, focus on clinical evidence*

Date: *December 2022*

Three Takeaways:

1. **Personal Experience Driving Innovation**: Sam's journey with ulcerative colitis and his personal challenges inspired the creation of Ayble Health. His firsthand experience highlights the importance of patient-driven innovation in developing effective solutions for GI conditions.
2. **Integrating Technology for Better Outcomes**: Ayble Health leverages machine learning and AI to personalize dietary and mental health therapy and increase access to multidisciplinary care teams. This technological integration aims to improve patient outcomes and make healthcare more accessible and effective.
3. **Commitment to Clinical Evidence and Collaboration**: Ayble Health emphasizes the importance of robust clinical research and collaboration within the healthcare ecosystem. By validating their innovations through rigorous trials and continuously evolving their standards

of care, they aim to set a new benchmark in digital GI healthcare.

Praveen: Sam, congratulations to you and your team for raising 4.6 million dollars. It's a fantastic start, and it's very reassuring for the industry and for digital GI. Let me start by asking you what the back story is here. How did you get started?

Sam Jactel: The story of Ayble is my story. I've been a GI patient for about ten years. I was diagnosed with ulcerative colitis, and I've lived with that. It hasn't been the most linear journey, and many other patients experience the same thing. I was diagnosed back in 2014, and I was misdiagnosed a couple of times before I got my final diagnosis. Over the past eight to ten years, I've seen five different gastroenterologists at some of the world's best academic medical centers. I've tried half a dozen different drugs, both oral and otherwise, over-the-counter supplements, and all sorts of things to find a cocktail that worked for me.

Despite having access to the best care, which is not the case for many people in the United States, I still fell out of remission. I had four significant flares in the last eight years, and I couldn't live like that anymore. One of the things I realized was that I was only really being treated as part of a person. I'd go into the clinic, and while drugs were available, they weren't always effective by themselves. My care team was helpful, but once I walked out of the clinic, I had no support for any of the evidence-based interventions my physicians recommended. These included elimination diets, dietary interventions generally, mental health support, cognitive behavioral therapy, and seeing a GI psychologist. The best I ever got was a pamphlet and instructions to figure that stuff out on my own. That just didn't seem good enough.

We were missing an opportunity to use new technology, like machine learning, predictive artificial intelligence, and other technical tools, to reduce the cognitive burden on the patient, improve the likelihood of achieving remission, and increase the accessibility of multidisciplinary care. That was our core thesis and the inception of the company. In early 2020, during a severe flare, I was bedridden and realized I couldn't do it anymore. I ended up working with a group of researchers at Massachusetts General Hospital, Brigham and Women's, and Northwestern to build a personalized nutrition program that worked for me. I realized I could systematize, standardize, and expand upon this program to help others.

Since 2020, we've been in the clinic running clinical research on this approach. Our Precision Nutrition program, our flagship program, is built on 20 years of research at the intersection of diet and IBS/IBD, as well as two and a half years of our own primary clinical research. This research has now been published in Gastroenterology, Inflammatory Bowel Diseases, and, most recently, Clinical and Translational Gastroenterology. We're now using some of the funding to commercialize this novel machine learning-guided Precision Nutrition program for folks like myself. At the end of the day, I solved my own problem and realized it could help others. That's where we came from.

Praveen: Let me ask you a very fundamental question before we get into the business stuff. You experienced GI as a patient. You went in and out of clinics, met different gastroenterologists, and so on. What's missing in the industry? You touched upon it, but I want to touch on the pain a little bit, if you don't mind.

Sam: How long do you have, my friend? Look, there's a lot that's going really well in the space. I serve on the Congressionally Directed Medical Research Program (CDMRP) as a patient reviewer for inflammatory bowel disease (IBD), and I'm seeing a lot of amazing research being done. There are new drugs and new approaches to team-based care for GI, and the folks in the space are coming around. It's a growing ecosystem that is evolving and doing better.

The problem I have is not necessarily a GI problem but a healthcare issue. The issue, and something I'm really passionate about, is that healthcare is the only market where the end user doesn't really have a voice. Here's what I mean by that. If you are a consumer of shoes, you go to a store and purchase a pair of Nike shoes you like. Nike will take that feedback and produce more of what people want. In healthcare, there's no direct way for someone like me to vote with my dollars.

What ends up happening is that the needs of the patient are either filtered or not heard. The patient's needs are filtered through the GI, then to the pharma company, to the health system, to the insurer, and it's like a game of telephone—it gets distorted. The challenge I have with GI and healthcare, in general, is that the companies building in the space—health systems, insurers, employers—are making decisions based on assumptions about what patients need.

We talk a lot about patient centricity, but unless you are a patient, there's always that little gap. Unless you've lived with it every day, you can't make assumptions about what it's like.

That's why you see shocking statistics about what it's like to suffer in silence. These conditions are taboo—we don't like to talk about poop. Patients like me are unproductive at work and uncomfortable, and some statistics show that patients like me are willing to give up 15 years of our remaining life expectancy for an immediate cure today. That's 20 percent of someone's lifespan—a significant need.

Patients have to be the CEO of their own condition because no one else is doing it for them. Patients are incredibly resourceful, building their own Excel models, writing in notebooks, and gathering information to problem-solve in an environment that isn't supportive. Being patient-centered is different from being patient-driven. Patients being founders is something I'm excited about in this space and others.

I've lived with this for ten years—ten years of market research. There's an opportunity to change the industry by putting the patient as the driver, not as the passenger.

Praveen: Let's talk a little bit more about the company. What is the operational model, how do you plan to get paid, and why did investors choose to invest in Ayble Health?

Sam: Yeah, it's an exciting place to be. Very concretely, Ayble is a precision health platform for folks with gastrointestinal conditions. I use the term "precision health" on purpose. In the space, you're probably familiar with precision medicine, where you take a lot of variables from an individual to predict what drug works best for that person. But what about everything beyond the pill?

We have built a comprehensive ecosystem of care beyond the pill, and it's built on the largest GI behavioral health database in the world. We're using machine learning and predictive AI to gather a bunch of that data and continue to use it. Eventually, we'll be in a situation where we can predict for each individual what the right pathway through diet and mental health will generate the most optimal outcomes in addition to the pill, not as a replacement. That's the ecosystem we built.

Our go-to-market strategy is very similar to companies that are category-defining in other places, like Headspace Health. It's essentially a B2C2B business model. One of the exciting things is that, because we've been patients, we know where they are, and we are going directly to consumers on purpose. This allows individuals to find evidence-based resources in addition to the work they're currently doing in the clinic and then transition that into the enterprise.

We sell through gastroenterologists, providers, and primary care groups as an extra tool in their toolbox and also through employers and insurers who want to drive down claim volumes and per capita claim costs for patients like me, who are incredibly expensive because we're incredibly sick.

Our approach is different; we're not a virtual or telehealth platform. We've built out a digital therapeutic that wraps around with hands-on care from a nutritionist, health coach, and dietitian. We have last-mile nutrition tools that allow us to suggest and find food products that are compatible with someone's diet. For example, if you can't eat onions, which are poorly digested for most people, it's hard because they're in everything and make food taste good. What if you're also

Halal, want to eat low sodium for other health reasons, and prefer organic food?

Praveen: Do the patients reach out to you directly, or do GI doctors, primary care providers, or anyone else refer members to you? How does the business model work?

Sam: For patients directly, it's on a subscription basis, like other health and wellness companies out there. For physicians, it's a referral pathway. We've built an entire ecosystem of tools for physicians. If you're a physician recommending Ayble to your patients, it's important for you to be in the driver's seat and understand how that data comes back to you.

For enterprise clients, it's similar to other groups with per member per month or per member per year payments. We're very confident in our data because we've published it to show cost reduction, and we're willing to go at risk with some of our fees to align incentives.

Praveen: What is the pricing model? Do the members pay a subscription?

Sam: Yes, exactly. The patients pay a subscription fee, which isn't much more expensive than Netflix but is much more effective. That's how we've orchestrated it. Enterprise contracts are relatively standard. It's important to pay for usage so we get paid for engaged members, which is crucial. We are also putting together a few contracts that are completely value-based, sharing in the dollar savings and improvement of outcomes for those individuals.

Praveen: There's a tendency in the digital health startup community to apply AI and machine learning to everything. I'd like to push back a bit. Is that really necessary? Isn't it more about not doing the wrong things versus doing specific things? For IBS, it's about knowing the triggers and managing them. Why is AI or machine learning required for this?

Sam: I love that question because those terms are often thrown around without real purpose. I'll answer this in two ways. First, I'll illustrate what it's like to do this on your own, especially around diet. Let's use a FODMAP pamphlet as an example. Patients are told to avoid FODMAPs, which are highly restrictive and hard to adhere to. They then have to reintroduce foods one at a time over a long period, which can take a year and a half. This process is complicated and burdensome, even with a dietitian's help.

Machine learning and AI reduce the cognitive load on patients, making it easier to gather and interpret data. This helps create a better, faster, more effective program for the next patient based on aggregated, de-identified data. We've shortened the experimentation process from a year and a half to 12 to 16 weeks, improving outcomes for 81 percent of IBS and IBD patients and achieving clinical remission for almost 70 percent.

Second, we're not just making care virtual; we're improving outcomes. Our goal is to continuously iterate and improve the gold standard of care using our data ecosystem. You can't achieve precision health without gathering and interpreting data with tools like ML and AI. While we may be incorrect sometimes, these models help us get better over time.

Praveen: Other companies in the wellness space are taking similar approaches for digestive diseases, and at least one has gone through the FDA with a digital therapeutics approach. Does the world need another digital GI company, or is this just a sign of the times? I'm asking this from a business perspective, considering both the clinical and patient sides.

Sam: It's a really important question. The reality is that GI is a gigantic market with unmet needs. We need as many people in the space trying different things to change the lives of patients like myself. Look at the diabetes market—there are half as many diabetes patients in the US as there are GI patients, yet there are hundreds of diabetes companies doing interesting work. They're all finding niches, generating returns for investors, and having sustainable business models. I think we'll see that in GI as well.

How you present yourself to the business ecosystem and your go-to-market strategy matters. The prescription digital therapeutics route is exciting, like what Pear Therapeutics is doing with substance use disorder and insomnia, and others like Mahana Therapeutics and MetaMe Health. They create a lot of value and capture almost all of it, similar to how drugs work. However, you won't get the volume needed with that model alone.

If 25 percent of a commercially insured population has a GI diagnosis and another 10 to 15 percent have undiagnosed or under-diagnosed conditions, there's a huge demand. Our goal is to make our science and innovations accessible to as many people as possible. GI conditions are complex and hard to manage with one tool alone.

There should be more collaboration in the system. These are "and" conditions rather than "or" conditions. You should do your diet, psychology, drugs, and physical activity. It needs to be tailored to the individual, and someone needs to quarterback that. In the absence of an ecosystem that makes this possible, we empower patients to do it and support physicians in extending that care.

Praveen: That was very well answered. Thank you for sharing that perspective. If there's a patient out there thinking, *I want to be in his shoes and take my problem head-on,* what advice would you have for them?

Sam: First, the reason we're called Ayble is that we want patients to feel seen because, a lot of times, we aren't. It's uncomfortable to talk about these issues, and we want to make you "ayble" to go to restaurants without fear, travel, and be defined by something other than your condition. To those patients, we see you; we've been in your shoes, and it's tough, but we're trying to solve what you live with daily.

For those thinking about being the CEO of their own company, it's a gift and a curse. The gift is that you have a unique understanding of the problem you're solving for your users, which is exciting because you know what to build and what not to build. The curse is "me-search" rather than research. My experience as a white, privileged, educated individual isn't universal. Assuming all customers are like you can lead to mistakes.

It's crucial to hear others and include diverse perspectives. We've talked to thousands of patients to inform our decisions.

It's not just about me. That's a key insight I've learned, and it's important to share.

Praveen: What's the future of GI and healthcare from the lens that you're seeing right now?

Sam: I won't beat a dead horse, but the future of healthcare is patient-driven. We often talk about being patient-centric, but it needs to be at the core of solving the actual problem. We can't get the outcomes we want unless the person who is sick gets better. Whoever better solves patients' needs will ultimately win in the market, not just for outcomes but also for traction.

We're very excited about running randomized control trials. We did three years of clinical research before taking a dollar from anyone. The bar for clinical evidence needs to keep going up. It can't just be about non-inferiority; the standard of care needs to evolve. We should be publishing more of our research in clinical journals, showing how we do it, and empowering others to keep evolving.

Currently, there's a big gap between the claims companies make in digital health, not just in GI but generally, and the evidence that supports it. Digital health companies need to run their own research to validate their innovations. It's challenging, expensive, and hard to do well, but science is crucial in this space. Leading with evidence drives trust, outcomes, and sustainability from a business standpoint. That's my challenge to the space, and it's something we're committed to as well.

Praveen: Sam, in closing, I want you to repeat that philosophical analogy you shared with me earlier about *being* the bat versus *knowing* the bat.

Sam: I'm glad we can nerd out about this. There's a phenomenal philosopher named Thomas Nagel. He has an amazing article in which he makes a proposition: Imagine you are the foremost scientist in the bat field. You know everything about the physiology of the bat, the neurons, the chemical signaling, and how the whole system works. You know how echolocation works and how they navigate their environment while being functionally blind. You can know everything about a bat, but Nagel's proposition is that even though you know all this, you will never know what it's like to be a bat. The "what it's like" aspect is super important in this space. The analogy connects to being a patient—understanding the experience from the inside. If we can think more like that, we can really change the world.

Praveen: Sam Jactel, I'm so glad that you came on the Scope Forward Show. I wish you and your team all the best, and I hope more patients like you step up, take action, and start companies in the GI space because there is a lot of need. Thank you once again.

Sam: Thank you so much for the opportunity. I welcome discussions with the GI community and others who are innovating in the space. It's an exciting time and the right place to be building for people in this ecosystem. Go forth and prosper. I appreciate the opportunity. Thank you.

Update from Sam (July 2024):

Ayble has significantly expanded the care we provide to patients and our reach across the United States. Our care platform now offers unlimited access to a GI-specialized multidisciplinary care team, a suite of AI-powered nutrition and psychology programs, and curated wellness tools to drive sustainable symptom improvement for patients across the entire spectrum of acuity. Currently, our services are available to millions of patients, and by 2025, we aim to reach tens of millions through partnerships with employers, health plans, and providers. We continue to lead with evidence-based practices and will publish our 15th peer-reviewed study by the end of 2024.

HYBRID GI CARE IS HERE TO STAY: SAM HOLLIDAY AND DR. TRETA PUROHIT OF OSHI HEALTH

Sam Holliday is the CEO of Oshi Health. His background is in engineering and healthcare management. Dr. Treta Purohit is a Medical Director for the company. She's also a practicing gastroenterologist. As a virtual GI care center, Oshi Health offers comprehensive digestive care through a multidisciplinary team of specialists, including diet and behavioral health support.

Themes: *hybrid care models, virtual GI care, value-based payment models, symptom control strategies, diet, behavioral health, access to care, patient satisfaction, collaboration between digital health startups and private practices*

Date: *April 2024*

Three Takeaways:

1. **Hybrid Care Models Improve Patient Outcomes and Access**: By offering multidisciplinary care through virtual platforms, patients receive comprehensive support that includes dietary, cognitive, and medical interventions, resulting in higher satisfaction rates and faster symptom control. This model also addresses the growing shortage of gastroenterologists, ensuring more patients receive the care they need more quickly.
2. **Innovative Diagnostics Will Transform GI Care**: Advances in technologies like liquid biopsy, stool-based testing, and pill robotics will revolutionize screening processes and early detection of GI conditions. These innovations will enable gastroenterologists to focus on high-yield procedures and provide more precise,

personalized care, ultimately improving patient outcomes and healthcare efficiency.

3. **Data-Driven Insights Enhance Care Delivery**: Leveraging data and advanced analytics, Oshi Health is able to track patient outcomes, identify effective interventions, and continuously refine care models. By generating robust evidence on the effectiveness of new therapies and diagnostics, virtual care models can contribute to the broader healthcare landscape and support the development of more effective GI treatments.

Praveen: Sam, the last time we had a conversation was right after the Series A funding of $23 million. Since then, you've raised a total of $60 million. The Series B was $30 million. Give us an update on what has happened since that point.

Sam Holliday: Thanks for having me back. It's been two and a half years since the first time I came on the show, and a lot has happened. At that point, we were pretty early in our journey. With that funding, we went on to run a clinical trial on our care model. We're a multidisciplinary care company in GI, so we provide dietary as well as cognitive, gut-brain interventions via licensed clinicians. We have nurse practitioners and physician assistants on our team as well who can support the medical side, and then great GI physicians like Treta who help oversee that team and make sure they guide our care.

We ran a clinical trial with a national health plan to study how our virtual GI care works. They were curious and wanted to know and measure our impact on a couple of things. Was it a good experience for patients? Was it contributing to positive clinical outcomes? Because we were focused on a commercially insured population, they wanted to know what the impact was on people's ability to work because their customers are employers. And we measured that, and we also measured the cost of care. We had really, really strong results. We ran that in three states, and what we saw was 98 percent satisfaction in our patients. We ended up with 92 percent of those patients achieving symptom control in under four months, which to us means the patient now understands what causes their symptoms to occur. And we've worked with them to implement treatments that they can now use to keep that down, whether that's medication, dietary change, or gut-brain intervention.

While we were running that trial, we started to get interest from the employer market to work directly with us. So, we ran three pilots with employers in partnership with Willis Towers Watson, one of the major consultants, to see how an employer could offer this as a benefit. Do we get the same results that we're seeing in the clinical trial? And we did.

Based on the results of the trial, we've now scaled into national contracts with health plans, as well as some of the regional health plans, like Blue Cross of Massachusetts, with whom we announced a contract. We have a national contract with Aetna that's got a value-based component. A lot of interesting things for us to catch up on today. When we chatted before, we were only delivering care in three states. We're in 25 now, and we'll be in all 50 states in the next six months.

(As of July 2024, Oshi Health is in 45 states)

Praveen: Awesome, and congratulations! Now, Dr. Treta Purohit, you work in the private practice GI world, and now you work in this virtual GI care model. What's the biggest difference that you see for similar types of patients?

Dr. Treta Purohit: This is a big dilemma that we have as private practice gastroenterologists. We have very high-volume practices. We see extremely complex patients, and even the most efficient and competent of us are stretched to our limits. The system is designed so that we see 15 to 20 patients a day. However, logistically and administratively, I cannot see that patient every week, even if I want to. That's when close clinical coordination starts falling through the cracks. Working with Oshi for a year, I realized what a huge difference this model brings because it's not tied down by the fee-for-service model, where you have to see a certain volume to meet a certain benchmark. Now comes this patient who has complex GI needs and wants to be heard, and their visits are 45 minutes to an hour long; that's unheard of when you talk about a traditional GI practice. I've seen these patients really get better starting from their first visit because they're getting heard. At Oshi, the secret sauce is that you have a registered dietitian and a behavioral health provider. You are getting focused brain-gut therapy. I do see these patients getting better faster with less testing and really leading their best lives because there's also a lot of time spent on educating them. Now they know what their disease process is.

It's the way GI was not delivered previously. When I wear the private practice hat and see a patient, I'm like, okay, this patient needs to go to Oshi. When I'm at Oshi, we see these patients who have complex GI conditions, and then they need to go back to the community for their procedures for acute inpatient in-person care. So, yes, it's two very different worlds that in an ideal world should coexist and not be separate.

Praveen: Are you able to provide any numbers that measure impact? It can be the commercial, economic, or clinical impact.

Sam: On average, we've seen, consistent with clinical trial results, $7000 to $10,000 of cost savings per patient compared to when they don't get the multidisciplinary support that we can provide. A lot of that comes from avoiding the ER escalations. We've seen stats that showed that GI patients are the number one driver of treatment and release ER visits. They go in, they're seen, they're worked up—often with imaging like CT scans—all racking up costs to the purchaser just to have the patient released with no intervention. Those are the avoidable episodes that we can prevent.

In our Aetna contract, we drove the screening colonoscopy quality measure from 78 percent at baseline to 95 percent. Those people needed to get in and be seen and screened for colorectal cancer, and we were able to coordinate that into the local market. That's another way we drive positive impact: high achievement of symptom control, quality of life improvement, and workplace productivity. We saw about a 50 percent improvement in workplace absenteeism, meaning people were missing half as many days. All these factors add to the cost and benefit story.

Praveen: What kind of clinical conditions are you typically seeing?

Dr. Purohit: We are seeing the whole breadth of GI conditions. Sometimes, we are seeing patients where we are the first medical provider they have ever seen. So, we are also making a lot of new diagnoses. A large majority of what we see is what is now called disorder of the gut-brain interaction (DGBIs), functional GI disorders. A lot of chronic abdominal pain, reflux, and IBS, which is very similar to a brick-and-mortar practice. It drives the high volume of care that's needed.

We are starting to see a good proportion of complex IBD patients. These are extremely complex patients, and when a community GI has that patient, they're trying their best to serve these patients. They require a lot of time.

We had a patient who was following up with their local GI but was struggling. He initially came to us for diet and behavioral health, and we started picking up some clinical acuity. That's where the clinical team jumped in and made some changes in management. Some of these are either doctor-shopping or stuck in a place where they're just going to fall through the cracks and end up in an ER.

So that's where we are collaborating very closely with the GI physicians when they're still driving the main care. It's very gentle redirection, but it's working as a team with the local GIs.

We have this weekly clinical meeting that is truly rigorous and evidence-based. Studies are being thrown around, and we are making very challenging decisions about patients, like what the right next step is, which will change treatment options, improve outcomes, and prevent unnecessary utilization. We respect the patients' time and resources.

Lastly, we are seeing a lot of healthcare maintenance. We will see a healthy patient, 60 years old, who has never seen a doctor. So then we say, hey, you need a screening colonoscopy, you're overweight, see the registered dietitian (RD), let's screen you for NASH. We are improving access and decreasing redundancy in the system so the patient doesn't have to be seen again to get their colonoscopy.

Praveen: In other words, Treta's colleagues from the private practice GI world needn't be worried that, hey, somebody's taking away my patients. This is a typical fear that the GI world would have with respect to anything digital from the outside.

Dr. Purohit: I love that you're addressing the elephant in the room. Right on. And I think with my Oshi hat and with my private practice hat, I've thought deeply about it. First, there is a dearth of gastroenterologists. By 2025, we're going to be short of 1,500 gastroenterologists. And if we keep getting on top of our screening colonoscopies, especially as the recommended screening age decreases, the number of patients who need access is going to grow while the number of gastroenterologists is shrinking. Praveen, you've talked about burnout. There's a high percentage of burnout. About 55 percent of the gastroenterologists are over the age of 55. So, the working population of GIs is shrinking.

We should focus on where we can deliver the most value. We get in the patients who need colonoscopies. We get in the pancreatic cancer patients who are waiting in line because ten functional dyspeptic patients won't get off the phone call because they need to be seen. That's where this hybrid role comes in, where Oshi can take this time burden off our plates. They can take these patients who have high requirements, need to see RD, behavioral health (BH), and need phone calls, and Oshi is perfectly designed for them. Meanwhile, our schedules open up for patients who have a positive FIT test and a family history of colon cancer and they're waiting six months to get a colonoscopy.

I truly feel that this hybrid model is going to coexist in the future, and it is the only way to deliver care. If there's a patient who doesn't have a GI for 100 miles, the only way he will make it to your office is if he sees an NP at Oshi who says, "Hey, let's go get your colonoscopy." We will be very cohesive in the future. As Oshi scales, the number of patients we see scales, and it also is going to funnel more patients and procedures to the offices. It will extend the reach of a brick-and-mortar office.

Sam: I'll add. People should get colonoscopies—it's the gold standard. There are resistant patients, and we share evidence, explain sensitivity and specificity, and make a strong case for colonoscopy when medically appropriate. If they refuse, we ensure they get a FIT or a Cologuard. We always recommend colonoscopy and navigate people into the local market. That's how we've increased the screening rate from 78 percent to 95 percent in one of our contracts.

Praveen: Thank you for clarifying that. How does Oshi Health make money? What are your revenue streams?

Sam: Our mission is to get health plans and employers to cover our care. We currently serve only commercially insured patients and have negotiated contracts with health plans and employers. We get paid when a patient comes to Oshi and at milestones as they go through our multidisciplinary care model. Patients meet every two to three weeks with a nurse practitioner or physician assistant as well as a dietitian and/

or GI psychologist to identify and address their symptoms. We get paid through our insurance or employer contracts, so there's no charge to our local partners. For example, a GI physician told me they needed to focus on colonoscopy and endoscopy due to high demand. By taking on the follow-up care for functional gut-brain patients, we free up their time for higher acuity care.

Regarding numbers, when we last spoke, we had zero revenue. Now, we are in the eight figures annually. We saw almost as many new patients in the first quarter of this year as we did all of last year. This year, we expect to see between 15,000 and 20,000 new patients nationwide. We have over 50 clinicians on our team and are continually adding more to keep up with the growing volume.

Praveen: Congratulations on that growth. That's really exciting because it's indicative of the trajectory and direction of what patients are seeking and where everything is going. Since you started, other virtual GI care companies have emerged, taking slightly different approaches but still in your space. How do you differ from competition?

Sam: We believe it's crucial to be able to make or adjust a diagnosis, flag potential underlying issues, order diagnostic tests, and prescribe medications. We're the only ones today delivering comprehensive medical care with multidisciplinary components. The app-plus-content-plus-coach companies can't do this. Their approach often results in more questions directed back to local gastroenterologists, generating more follow-up care requests without relieving the pressure on practices. We've heard from practices that these models just generate more calls and pings for follow-up care, which isn't reimbursed at the same rate as new patient visits or procedures.

Some companies that tried these models have gone out of business. We've also heard skepticism from health plans about app-only approaches or content-plus-coaching models that lack clinical components. We've differentiated ourselves by being the only company with a clinical trial run by a health plan, demonstrating cost savings with a control group.

Praveen: Treta, you've been part of the GI Mastermind program, and we talk a lot about exponentials in the program. I'd love to know how you see Oshi as an exponential care model from which the industry at large could learn.

Dr. Purohit: Oshi is the perfect example of addressing a significant gap in the industry—providing multidisciplinary care to patients with chronic GI conditions. Traditionally, this level of care was only accessible at academic centers. Now, with Oshi, there's a digital solution that patients can access via their phones. It's a patient-friendly technology providing access to a comprehensive GI team, including highly vetted physicians, nurse practitioners, physician assistants, dietitians, and behavioral health providers, all working together.

With technology, patients across the country can access these resources regardless of location. It's like democratizing the care that used to be confined to academic centers. This is truly exponential. For example, a GI in the community can refer a patient to Oshi digitally, and a patient in Oregon can see a dietitian in New York who specializes in IBS. This is the power of exponential technology—it's not limited by time, place, or position. It's creating abundance and increasing access through employer contracts.

GI is complex, involving the gut-brain axis and many nuances that require comprehensive care. Simple algorithms or devices have value but can't provide the holistic care that many GI patients need. That's the value Oshi has created—offering a complete, digital care model that is poised to become a standard referral option for GIs everywhere.

Sam: If I can add one piece on top of that, we're also tracking a tremendous amount of data about the patients we're serving. As the world continues to evolve with AI and machine learning, that data set will help us figure out which interventions work for which types of patients. This will make us quicker and more efficient at achieving outcomes for patients. It also sets us up as a research site for new diagnostics or therapeutic interventions, contributing to generating evidence on new approaches.

Praveen: As we wrap up our conversation, Treta and Sam, I have a question that I ask everybody, and I'd love your individual reflections on this. Where do you see the future of GI going?

Dr. Purohit: The digital wave is here to stay, and it's going to bring several changes. Digital health and hybrid care will become the standard because they improve access and address the issue of fewer gastroenterologists. While we love doing our procedures, there aren't enough gastroenterologists to screen everyone. We'll see a rise in non-invasive stool-based testing and advancements in liquid biopsy and other less invasive diagnostic procedures - lucid diagnostics and pill robotics.

The diagnostic space will transform, allowing gastroenterologists to focus more on high-yield procedures and patients who need more attention.

Sam: From our perspective, we're working to ensure coverage for our services is ubiquitous, making it easier for local gastroenterologists to refer patients to Oshi without worrying about coverage. When we achieve this, there will be a seamless exchange between local gastroenterologists and Oshi. This will strengthen partnerships and expand the reach and scope of practices, allowing them to serve every patient and retain those relationships.

Unfortunately, current satisfaction rates with GI care are low. We've measured it in our clinical trials and have seen surveys showing only 29 to 38 percent of patients are satisfied with their care. This can improve as we solve underlying problems for patients together. When patients view their local GI and Oshi as a seamless partnership, satisfaction rates will rise to 100 percent.

Praveen: Sam and Treta, this has been a fantastic conversation.

WHAT IF WE DIDN'T DO AS MANY COLONOSCOPIES? DR. MICHAEL OWENS OF PEARL HEALTH PARTNERS

Dr. Michael Owens is the cofounder and partner at Pearl Health Partners. He's a practicing gastroenterologist. Pearl Health Partners offers value-based outpatient surgeries and procedures, providing high-quality care from top physicians at lower costs.

Themes: *multispecialty collaboration, value-based care, autonomy in healthcare, innovation in private practice models, pricing transparency, independent practice*

Date: *October 2022*

Three Takeaways:

1. **Innovative Multispecialty Approach with Value-Based Care:** Dr. Michael Owens transitioned from private practice GI to founding Pearl Health Partners, a multispecialty group focusing on value-based care. This model allows for comprehensive and integrated care across various specialties, leveraging the benefits of multidisciplinary collaboration to improve patient outcomes and reduce costs.
2. **Redesigning GI Procedures and Emphasizing New Technologies:** Dr. Owens and his team have rethought the approach to GI procedures, such as colonoscopy. Innovations include same-day preps, piloting new food-based prep solutions, and exploring hydrotherapy for colon preps. They are also considering single-use devices and emphasizing microbiome research, which could significantly impact future GI care practices.

3. **Autonomy and Strategic Planning in Private Practice:** Dr. Owens advises maintaining autonomy and ownership among care providers when considering bringing on capital or liquidity events. He highlights the importance of strategic planning and the benefits of remaining independent from private equity and larger hospital groups, allowing for greater flexibility and focus on patient-centric innovations and care models.

Praveen: Dr. Michael Owens, you've had a long career in interventional gastroenterology, and now you've built Pearl Health Partners, an independent organization focused on value-based care with two surgery centers and multidisciplinary specialists. You're expanding in the Portland metro area. It's fascinating that you transitioned from a private practice GI to this multispecialty GI entity. Please share more.

Dr. Michael Owens: Thanks for having me, Praveen. We have had some interesting conversations already. Reading your book, "Scope Forward," I realized there were others out there with similar thoughts, especially pre-pandemic. In the past four or five years, there's been increasing pressure on revenue models and the healthcare system. Larger organizations meant more complexity, and the independent practice started to feel the strain.

I spent almost 20 years as a therapeutic endoscopist, working three days a week at the hospital, building service lines, and helping to establish an advanced endoscopy center. By 2021, I saw opportunities to take control of our situation. We spent six months developing the idea, made an offer on an unused surgery center, and quickly expanded. We brought in surgeons and other specialists, focusing on service lines like pelvic floor dysfunction and weight loss.

I started thinking, *What would I do with my time if I'm not doing as many colonoscopies?* It was refreshing, like a thought experiment. I was excited about the microbiome and weight loss, areas where we, as GI doctors, had often failed to spend adequate time. Many people considered weight loss a primary care diagnosis and treatment algorithm, but I saw it differently. I was interested in working on service lines with other doctors to whom I used to send patients, and it would take forever for them to be seen because they were in different institutions.

We aimed to build a new way of working together centered around a surgery center. We saw hospitals struggling to provide access to care and realized we could offer a better alternative. My cofounder, Dr. Richard Rosenfield, is a GYN surgeon with extensive experience in the business of medicine. Together, we explored value-based care and transparency, partnering with organizations for data mining and market analysis. We believe we can deliver high-quality care at a lower cost, benefiting both patients and payers.

Our goal is to create a sustainable, independent group with enough autonomy and equity. We want to avoid being just another acquisition target for PE firms. We're focused on multispecialty care, incorporating fields like cardiology, to ensure we stay ahead of changing healthcare dynamics. We've consulted with experts to guide us, and we're determined to execute our vision while retaining our independence.

Praveen: Let me clarify a few things. You didn't go for a single GI specialty type expansion; you went the multispecialty route. I don't know of any other gastroenterologist in the country who has dropped out of private practice and started a multispecialty group with a focus on value-based care. Congratulations on that. How is this all funded? Is it self-funded by the physicians, or did you raise money?

Dr. Owens: We were in a unique position where our cofounder had an ASC that mostly did GYN plastics and some other specialties over the years. It was already capitalized, recently recertified, and had established contracts. We needed to renegotiate some contracts due to the new additions, but we could bring in investors at a low EBITDA. The goal is to allow more people to get involved without a high barrier to entry, enabling them to start working and generating revenue quickly.

The physicians are funding the practice and the remodel of a second center with their equity shares. This model is tailored to our specific situation. It wouldn't be a copy-and-paste solution for other regions. Starting from scratch in a new suburb with a $5 million investment in an ASC would require different capital needs. We just received confirmation today that we can expand to our second center under the same tax ID with Medicare, which will help us avoid delays.

Our process involved a lengthy setup for articles of incorporation and the operating agreement to create this specific version of our model.

Praveen: Got it. I'm assuming your partners were part of other practices. Were they full-time with this venture, or is it part-time while still engaged in other practices?

Dr. Owens: They are all from independent practices. We have nine different practices now, and each has its own components. They're interested in eventually bringing all those practices under one roof for the benefits it offers. Our steps included one surgery center that's already operational but not big enough for all of us, so we're expanding. Combining our practices brings leverage for contracting and transparency data. It's been almost entirely about group dynamics and opportunities. Everyone involved is innovative and sees the value in maintaining independent practices as long as possible.

Praveen: Like you, your partners didn't want to join private equity groups or larger hospital groups?

Dr. Owens: Exactly. They're committed to maintaining autonomy and independent practices.

Praveen: Beyond the business aspect of coming together in a multispecialty environment, are there any correlations you're seeing in patient care itself between GI and other specialties?

Dr. Owens: Absolutely. We've seen clear connections, especially with women's care. Among a few practices, there are around 10,000 patient lives, and fatty liver disease is a significant issue. We found immediate alignment on issues like pelvic floor endometriosis, where our colorectal surgeon, pain specialist, GYN, and GI practices overlap, especially with microbiome work. We haven't focused yet on complex IBD or liver disease since we started with outpatient procedural and surgical care, where our business model evolved. Surprisingly, the clinical care alignment happened naturally.

For example, our GYN surgeon recently consulted with one of our foregut surgeons about a patient. Such interactions are common in larger multispecialty groups but differ here because we control our processes. We all aim to deliver value-based care and support the center financially. From a procedural aspect, it's easier to bundle costs and figure out pricing. It's more complex from a payer's or hospital's perspective when looking at the entire service line as a silo. Different inputs and outputs shape those thought processes.

Praveen: So you've talked about pricing transparency and negotiating with insurance companies. Is this model better suited for negotiating than a single specialty GI model?

Dr. Owens: From the discussions we've had at a system level with administrators and financial officers, they find this type of innovative thinking very attractive. It's difficult to say if it works universally, but they see strategic value in our approach. We recognize there are moving pieces and challenges. For example, we haven't yet built a program for Crohn's patients, but we'll get there. Adding specialties that aren't proceduralists, like endocrinology for our weight loss program, will require careful integration. Existing multispecialty groups have their own ways of managing ASCs and finances, so there are trade-offs. Ideally, the US healthcare system would shift away from fee-for-service, solving many issues, but in the meantime, we're focusing on what we can achieve now.

Praveen: Can you outline what innovations are happening in your business that are different from those of other multispecialty or private practice companies?

Dr. Owens: Bringing together all these specialists under one tax ID and working together to deliver care at a lower cost are parts of the biggest innovation. For GI-related areas, I got to rethink colonoscopy from the ground up. We decided to do things a little differently. We don't have like a group of 60 doctors with three committees taking six months to pick which rep we use. So, there are benefits, right?

We rejiggered our day around our ASC because there are not as many people needing to get in to get their cases done. We're doing mostly same-day preps, about five hours before the two-hour NPO window, and it's been going great. Patients love it. I love it. They're not up all night suffering and calling me. Now, in the morning, we do uppers, diabetics, and constipation patients, so we've pieced that together differently.

We're piloting the Happy Colon low-residue food prep for people who don't want clear liquids, which has been more accepted. High GI care is finally more accepted, too, and we're looking into hydrotherapy for colon preps. Portland has experts at colonics, and it's fascinating. We used to be cautious, but there are people doing it right. We're getting data now from large PE firms and GI groups that are building centers for IG care.

Our microbiome work is interesting. The effluent from a colonic has different microbiome diversity than spontaneously passed stool. We thought about what we could do differently for patients and procedures. We're looking at single-use devices and waiting for prototypes and pricing to decide if that's where we'll head. We don't want to invest heavily in equipment if colonoscopy volumes drop significantly in the coming years.

Quality programs are a priority. We've done all the American Society for Gastrointestinal Endoscopy (ASGE) quality programs to maintain our data. We communicate clearly with patients about why they should have their first colonoscopy at specific ages, considering alternatives like Cologuard. There are many companies working on non-invasive approaches, and the science is real.

Risk stratification is key. AI can double the ability to detect small polyps, but it changes the risk category. New tests with better detection don't mean high risk at lower thresholds. We're seeing more people become aware of alternatives, and Exact Sciences like the Blue Sea trial are leading the way.

For comprehensive care, we see a lot of alignment, especially with women's clinics and pelvic floor dysfunction. Fatty liver, pelvic floor issues, and endometriosis align with colorectal surgery, pain specialists, and microbiome care. We're building around outpatient procedural and surgical care first, but complex IBD and liver disease programs will come later.

Transparency and negotiating with insurance companies are easier with our model. It's attractive to health systems and health plans. We're integrating transparency data and building new relationships with organizations doing deep data mining. We're looking at how to bundle services and make healthcare more affordable.

Bringing in endocrinology for our comprehensive weight loss program is another step. We're figuring out the pieces as we go, but our approach has been to deliver high-quality care at lower costs, maintaining independence and autonomy. It's a different model, but we believe it's the future.

Praveen: It's concerning. It's amazing. What changes do you see in private practice, gastroenterology, or GI as a space as a whole five years from now?

Dr. Owens: Well, if I were to follow what Dr. Lawrence Kosinski said recently, pointing out that people have their heads in the sand with colonoscopy factories, I'd wonder about five years from now. I'd wonder what we will do with all those GI rooms and what half of my partners will be doing all day. One thing that makes me feel this model we're building is interesting is that I have seven people doing endoscopy and colonoscopy, but half of them also do surgery. We used to look at that as competition and turf battles, but training is different now. The trainees coming out are very talented endoscopists who can do a lot more than I can as a GI doctor. They have skills I don't have, which can be complementary.

We'll have a GI team that's advanced in microbiome and Crohn's care. We're not going to ignore it anymore. There's a probiotic mix showing benefits in mouse models. The data is sound, and therapies and monitoring will finally address weight loss, not just genomics. There are phenotype variants between nature and nurture that we're starting to figure out.

Therapeutic endoscopy will become more specialized, working with a lot of professionals who aren't just traditional gastroenterologists. AI will sort out a lot of the complexity for us. We won't be finding insignificant polyps because panels from multiomics and the microbiome will tell us this person doesn't need a colonoscopy. We'll be able to change diets and improve health markers like bacteria levels.

We might spend more time with patients and be comfortable with less revenue because our quality of life and intellectual stimulation will be higher. We'll have fewer repetitive motion injuries. These are all very possible. Will there still be independent practices? I don't know.

Praveen: Yeah, there will always be the old with the new because some aspects of any industry won't change or will take a long time to change. But in closing, Mike, any final words of wisdom for people who might be in your shoes from a couple of years ago, contemplating their next steps and deciding what to do?

Dr. Owens: I've had these conversations lately. If you're bringing on capital, considering a liquidity event, or bringing in non-clinical partners, think hard about that. You can, and maybe should, maintain ownership among the people providing the care. Carefully consider the value of large sales where you lose autonomy and what that really means. PE firms and VCs are thinking along those lines, too. If your group decides to go down that path, maybe for the typical reasons of exit strategies for older docs, look at what happens down the road. Start considering other models where you and a few aligned colleagues in your gastroenterology group think about what else you could do. Try not to give away the whole boat, and talk to those you think are aligned about what to do next.

Praveen: Mike, this was fantastic. The way you've outlined and connected the dots, I'm sure many will benefit from it. Thank you.

Dr. Owens: Praveen, thank you so much. It was a pleasure.

BRIDGING THE BRAIN-GUT CONNECTION: DR. MEGAN RIEHL AND DR. MADISON SIMONS

Dr. Megan Riehl and Dr. Madison Simons specialize in psychogastroenterology. Dr. Riehl is the coauthor of Mind Your Gut: The Science-based, Whole-body Guide to Living Well with IBS and cohost of The Gut Health Podcast. Dr. Simons is a psychologist specializing in gastroenterology, hepatology, and nutrition at the Cleveland Clinic.

Themes: *brain-gut behavior therapies, psychogastroenterology, cognitive behavior therapy, IBS, IBD, Rome Foundation, digital therapeutics*

Date: *June 2022*

Three Takeaways:

1. **The Role and Impact of Psychogastroenterology:** It's a field that has evolved over the last couple of decades, focusing on the application of psychological interventions for treating gastrointestinal issues. These interventions, known as brain-gut behavioral therapies, have shown effectiveness across various GI conditions, including irritable bowel syndrome, inflammatory bowel disease, and esophageal conditions.
2. **Patient and Provider Demand for GI Behavioral Health:** There's a high demand for GI behavioral health services, particularly in tertiary care clinics. Collaborative relationships between GI psychologists and referring providers enhance the overall treatment plan and patient care.
3. **Future Directions and Innovations in GI Behavioral Health:** The future of psychogastroenterology includes developing triaging models to prioritize patient care and

exploring the metabolic consequences of dietary modifications. Innovations like digital therapeutics will play a significant role in providing sustainable resources for managing conditions like IBS and IBD.

Praveen: I'm really glad to have this conversation. To start, what is psychogastroenterology? It sounds very cool, but what is it?

Dr. Megan Riehl: It's really great to be here and have this conversation with Dr. Simons and you, Praveen. Psychogastroenterology is a field that has really evolved over the last couple of decades, but we've seen an emergence of exciting research around the application of psychological interventions for the treatment of gastrointestinal issues. We call these interventions brain-gut behavioral therapies. It encompasses the field of psychogastroenterology, where we work as expert GI psychologists, fitting into a multidisciplinary approach with patients with a variety of GI conditions. The bulk of the research has been done in the functional GI world or disorders of gut-brain interaction, but we have exciting research happening in inflammatory bowel disease and esophageal conditions. We're finding that our treatments are effective no matter where they fall in the gastrointestinal tract.

Praveen: Does it have to be very specific to the GI tract? Could it not be broader psychology that might also apply to the GI tract?

Dr. Madison Simons: Absolutely. It could be underlying mental health conditions like anxiety and depression that could be exacerbating the GI symptoms. But I'm sure Dr. Riehl can attest to this as well: if the underlying mental health condition is very severe, then our specific GI interventions are not going to be as helpful. In that case, we would pull on a colleague who is trained to deal with anxiety or depression to address that first and stabilize it so that our GI interventions can be really targeted to the GI tract. The patient demand for our services is so high just in addressing GI symptoms that we're better suited to use our expertise for the GI symptoms and allow other colleagues specialized in anxiety, depression, and trauma to treat those conditions.

Praveen: What does a typical treatment plan look like in psychogastroenterology? How do you incorporate these brain-gut behavioral therapies?

Dr. Riehl: A typical treatment plan involves several components. We often start with psychoeducation to help patients understand the brain-gut connection and how their mental health impacts their GI symptoms. Cognitive-behavioral therapy (CBT) is a cornerstone of our approach, helping patients change negative thought patterns and behaviors that may be contributing to their symptoms. We also use techniques like gut-directed hypnotherapy, mindfulness-based stress reduction, and relaxation training. These therapies are tailored to each patient's specific needs and conditions, and we work closely with gastroenterologists to ensure a comprehensive, integrated treatment plan.

Dr. Simons: In addition to these therapies, we also focus on helping patients develop coping strategies for managing their symptoms in daily life. This might include dietary modifications, stress management techniques, and lifestyle changes. By addressing both the psychological and physiological aspects of GI conditions, we can help patients achieve better overall health and improve their quality of life.

Praveen: You're saying there is a lot of patient demand for the services that you offer. Why and what kind of patient demand?

Dr. Riehl: We work in tertiary care clinics, so I have over 100 gastroenterologists and trainees who can refer to our behavioral health program. I started out as a team of one, and over the last couple of years, we've expanded to three full-time GI psychologists to meet the demand. When I started back in 2014 at the University of Michigan, once the program was established, providers began to see the value we bring in terms of reducing healthcare utilization and helping patients who have been suffering from symptoms for decades. These patients often haven't responded well to medication or diet therapy, but our behavioral interventions tend to be very effective for those refractory cases.

In a relatively short period of time, anywhere from five to seven sessions, patients show significant improvement. This success increases our referrals, and we form collaborative relationships with our referring providers, working from a multidisciplinary perspective. Patients appreciate that their gastroenterologist isn't expected to meet all their needs alone. However, we do run into wait times, so as Dr. Simons mentioned, ensuring that the appropriate patients are referred to us helps to avoid bottlenecks in our referral streams.

Praveen: What kind of patients or conditions are you seeing, Megan?

Dr. Riehl: We see a lot of patients with irritable bowel syndrome, inflammatory bowel disease, GERD, and various esophageal conditions. It's always rewarding to use our esophageal hypnosis protocols, which are really effective for functional dysphagia and globus. These are difficult-to-treat conditions where our behavioral interventions can make a significant difference. We also work with patients who have gastroparesis, though some of these cases may require referral or close collaboration within an integrative team. Essentially, for most GI conditions, we can add valuable components to the treatment plan that patients find beneficial.

Praveen: Megan, you've worked in the field for several years now. What are the insights you've taken away that have been surprising for you personally, having applied the field and seen so many patients? What have you personally taken away?

Dr. Riehl: One key insight I've gained is how powerful it is to explain visceral hypersensitivity and its symptoms to patients. Madison mentioned the reluctance some patients have when referred to a psychologist. Educating our referring providers on how to explain the referral is crucial. It's not about the symptoms being in their head or them being psychologically damaged; it's about managing their symptoms differently.

When patients understand brain-gut dysregulation, they realize their symptoms are real but are due to miscommunication between the brain and gut. This explanation can be eye-opening for them. We focus on decreasing the awareness of uncomfortable sensations using tools and strategies, aiming to reduce the frequency, duration, and severity of symptoms in a short time.

Watching evidence-based interventions repeatedly work is incredibly rewarding. Even if symptoms aren't completely alleviated, patients often gain confidence in managing their symptoms. It's fulfilling to hear patients say they wish they had this treatment 30 years ago or to see young patients learning strategies that will help them lead healthier lives.

Praveen: Are gastroenterologists open to it as well?

Dr. Riehl: Definitely. When I first started in Michigan, they had never had a GI psychologist, and many places still don't have one. Both Dr. Simons and I trained at Northwestern University, where there was an established program, and I took that model to Michigan. Initially, it's about program development, building a business model, and encouraging referrals. It really is like a "field of dreams"—if you build it, they will come. Once a patient has a positive outcome and the referring gastroenterologist sees fewer messages in their inbox, it creates buy-in.

What's been fun and fascinating is our fellowship program. Our fellows come in with a multidisciplinary approach, referring patients to GI dietitians, behavioral therapists, and physical therapists. When they move on to different places, they realize how crucial this multidisciplinary team is. They might come back saying, "There's no psychologist where I am," and then they advocate for building a behavioral health program in their new hospital or clinic, especially if they're motility specialists.

Praveen: How many GI psychologists are there in the country?

Dr. Riehl: Just over 400 worldwide. The Rome Foundation has a special subsection of psychogastroenterology, and they have a provider directory that vets members based on expertise. There are just over 400 members around the world, which isn't nearly enough considering there are 40 million Americans with IBS alone, not to mention other GI conditions.

Praveen: Where is the field going? What do you anticipate seeing in the future?

Dr. Simons: The biggest challenge is that there are so few of us. At Cleveland Clinic, I'm working on developing a triaging model to prioritize the most critical patients. We need to explore different ways to provide services. My interest is in dietary modification. Almost every patient I see has changed their diet in some way, and I'm interested in the metabolic consequences of that. For instance, how do changes in diet affect blood sugar and glycemic variability in inflammatory bowel disease? We need to look beyond just weight loss or nutritional deficiencies. I hope to create treatments specifically for anxiety and fear around eating.

Dr. Riehl: The innovation of digital therapeutics is happening now. There's a lot of work to do to educate both patients and providers on how to utilize these tools effectively. Those of us seeing patients in person or virtually will still have plenty to do. Marrying digital tools with traditional care can help sustain lifestyle changes for managing conditions like IBS, IBD, or gastroparesis. With changing insurance billing models, having tangible, sustainable resources that patients value is essential. A GI psychologist can play a crucial role in delivering these outcomes. The role of the GI psychologist in direct patient care, research, and leadership is vast and evolving.

Praveen: Any final words, Madison, from you and Megan?

Dr. Simons: Absolutely. Being on the ground, working with gastroenterologists, having come from Northwestern, and now starting at Cleveland Clinic, I hear every day about the value I bring to the team as a GI psychologist. For private practice gastroenterologists, there's no question about the benefits of hiring a GI psychologist. From a financial perspective, the patient treatment burden and helping clinicians understand and develop empathy for their patient's symptoms are of immense value. It's incredibly rewarding for me and the team I work with. Any provider would benefit from including a GI psychologist in their practice.

Dr. Riehl: Madison and I both have extensive training as GI psychologists, but that level of training may not be necessary for every mental health provider interested in working with patients with IBS or IBD. Even if you're a gastroenterologist or a primary care physician, there are opportunities for general mental health providers to receive additional training through the Rome Foundation, which offers excellent continuing education resources. We need to think creatively about collaborative relationships between gastroenterology practices and mental health. If you can't find a GI psychologist like Madison or me, forming strong relationships with local mental health providers can still bring significant value. If a mental health provider enjoys working with GI patients, the Rome Foundation's resources can help them become more proficient. This collaboration can be a win-win for your practice and for the mental health provider, providing a steady stream of referrals. There are many ways to foster these beneficial partnerships.

VIRTUAL REALITY THERAPEUTICS: DR. BRENNAN SPIEGEL, AUTHOR OF VRX

Dr. Brennan Spiegel is the Director of Health Services Research at Cedars-Sinai Health System and author of *VRx: How Virtual Therapeutics Will Revolutionize Medicine*.

Themes: *virtual reality, pain management, IBS, neurohormonal pathways, non-pharmacological treatments, brain-gut axis, cognitive restructuring*

Date: *March 2021*

Three Takeaways:

1. **VR as a Therapeutic Tool**: Virtual Reality (VR) is emerging as a significant therapeutic tool for various medical conditions, including pain management, anxiety, depression, stroke rehabilitation, and gastrointestinal disorders. Its ability to distract the brain, promote cognitive restructuring, and facilitate a calming state provides substantial benefits for patients with chronic conditions like IBS.
2. **Integration into GI Practices**: For busy gastroenterologists, incorporating VR into their practice can offer an effective way to manage conditions like IBS without significantly increasing their workload. VR programs can provide patients with skills for managing their symptoms through cognitive behavioral therapy (CBT) and mindful meditation, potentially reducing their reliance on medications and frequent hospital visits.
3. **Future of VR in Healthcare**: The mainstream adoption of VR in healthcare is likely to accelerate with increasing evidence of its efficacy, insurance coverage, and clinician acceptance. As patient demand grows and VR technology

becomes more accessible and integrated with other digital health tools, it will play a crucial role in revolutionizing patient care and expanding the scope of treatment options available to healthcare providers.

Praveen: Dr. Brennan Spiegel, my first question is about your author photo on the cover of your book *VRx: How Virtual Therapeutics Will Revolutionize Medicine*. The same picture appears on Cedar Sinai's website where you work. Something significant must have happened that day with that patient. Can you share more?

Dr. Brennan Spiegel: No one's asked that question before. That photo was taken several years ago when we were first testing virtual reality (VR) with patients. For those unfamiliar, VR is often associated with gaming or entertainment. About six years ago, we started exploring its use in hospitals for pain management. The day the photo was taken, we were working with a young woman who had severe recurrent abdominal pain. She had been hospitalized six times in one year and tried various medications, including opioids and ketamine, without much relief. We decided to try virtual reality. When that photo was taken, I was reacting to her response to the VR. She went from being frustrated and upset to being immersed in a virtual world, reaching out to blue whales and laughing. She was genuinely enjoying herself, and I couldn't help but smile in response. We had a photographer present that day, and I've used that photo as my headshot ever since because it captures a genuine moment of a patient benefiting from VR.

Praveen: Yes, and that comes through in that photograph. Brennan, why did you choose to use VR that day? I'm sure you encountered many patients with abdominal pain around that time.

Dr. Spiegel: In gastroenterology, we all learn about the brain-gut axis. The brain and the gut are connected, which makes sense. The old notion from René Descartes in the 1600s suggested dualism and that the brain and body are separate. However, we know that's false. The brain is intricately tied to the gut and vice versa through neurohormonal pathways. Cognitive behavioral therapy, psychotherapy, hypnotherapy, and other non-pharmacological treatments have long been known to help people with IBS. These treatments aren't always cures, nor are they replacements for traditional medical therapies, but they can support people with IBS.

With that background, when I learned about VR and its ability to positively influence the human mind, I thought, why not try it with people with IBS, especially those in the hospital where our treatments are often inadequate? When I started to see positive responses, I realized we were onto something worth studying. Six years later, we've used VR with over 3,000 people at Cedars-Sinai and learned a lot, which we can discuss further today.

Praveen: What are three or five takeaways from your experience with VR?

Dr. Spiegel: I'll start with what VR is doing and how it's working. Clinicians want to understand the mechanism of action for any new treatment, whether it's a drug or a non-pharmacological therapy. So, the first takeaway is understanding the mechanism, particularly for pain. But it's important to note that VR has been used for many conditions beyond pain, including eating disorders, anorexia, obesity, dementia, schizophrenia, anxiety, depression, stroke rehabilitation, autism, cerebral palsy, multiple sclerosis, and more. There are over 5,000 studies supporting VR's efficacy across about 50-60 conditions.

For pain, one question is, how does VR help? There are different mechanisms. One is distraction, known as inattentional blindness. Humans can't concentrate on many things at once. VR draws attention away from nociceptive experiences like pain. Additionally, VR seems to help the brain fight back against pain. When the brain is in a calm state, it can inhibit pain signals through descending inhibitory pathways. VR induces a state of mind that helps the brain inhibit pain using the gait control theory, where the brain sends inhibitory signals down the spinal cord to close pain gates.

Another mechanism is cognitive restructuring, which helps people rethink their pain. This is crucial for chronic pain, especially visceral pain like irritable bowel syndrome and functional abdominal pain, where medications often fall short. Cognitive behavioral therapy aims to help patients understand the relationship between their pain, body, and mind, and VR can facilitate this process.

Praveen: Great. I want to ask you about the story of these patients after the first few visits. Someone has IBS, and they're experiencing something new, like dolphins swimming or a surreal trip from the VR app. Initially, it's new and engaging, but what happens over time? Does their reliance on drugs reduce? What happens three months, six months, one year, or even two to three years down the line?

Dr. Spiegel: We're starting to see studies with longer-term follow-up, though not yet in GI, but in other areas like chronic lower back pain. A recent randomized control trial for chronic lower back pain used an eight-week skills-based CBT treatment in VR compared to a sham VR, where the control group watched neutral, two-dimensional scenes. They followed patients for eight weeks. While we don't have one year of data yet, we do have eight weeks of data showing that the separation in pain relief between the VR group and the control group not only persisted but grew over the treatment period. By the end, patients in the VR group had significantly lower pain levels, both statistically and clinically.

The studies aim to reduce the need for prolonged VR use. Instead of increasing VR usage, the goal is to teach patients skills in VR that they can apply in their real lives. This helps them enjoy real reality (RR) without relying on VR. VR is not meant to be an addictive tool like a game but a means to teach valuable skills for managing their conditions.

In GI, we're creating a comprehensive IBS VR program and intend to test it over longer periods to gather more data. The aim is to see if the benefits of VR for managing IBS and other GI conditions hold up over time, reducing reliance on medications and improving overall quality of life.

Praveen: Excellent. So, in your book VRX, Brennan, I remember a story about a patient with abdominal pain who experienced VR. Despite trying everything else, nothing worked until that moment. She realized her abdominal pain was linked to her brother's death from stomach cancer. Can you talk about some of this story? I'm sure the GI community sees similar patients, but it might not make these connections as you did in that example.

Dr. Spiegel: Yes, that's a powerful story I often share. This patient had recurrent severe abdominal pain of unknown origin. She had undergone a comprehensive workup, including upper endoscopy, colonoscopy, CT scans, and lab tests for various conditions. Despite all this, we couldn't pinpoint the cause of her pain. She was in the hospital, and we were puzzled.

We decided to try virtual reality. I used a headset to put her in a scene where she was swimming with dolphins, a pleasant and calming experience. She became silent for about four minutes and then started to cry. When I asked if she was okay, she said she thought she knew why she had the pain. She linked it to her brother's death from stomach cancer and her fear of having the same fate.

Even though we had found no signs of cancer in her tests, she said she hadn't been able to accept that until the VR experience. She felt the dolphins were telling her to accept reality and move on with her life. Remarkably, she reported that her stomach pain was gone, and she felt ready to go home.

This experience was incredible because it showed how VR could help someone make an emotional and psychological breakthrough. It doesn't work this dramatically every time, but for her, it was the right trigger to reboot her brain. Studies with MRI scans show that VR can power down the brain's default mode network, which controls our inner voice, allowing for lateral thinking. This is similar to the effects of meditation and psychedelics. We sometimes call this a "cyberdelic" experience.

Praveen: A few years back, I was in Peru with a native shaman and went through an ayahuasca ceremony. It was semi-psychedelic, so I experienced firsthand what happens. When I see VR apps like Tripp or ones that replicate these experiences, it seems that calling VR a technology tool and doing clinical studies gives credibility to ancient healing modalities. The healthcare system seems more open to this now. I'm curious about your comments.

Dr. Spiegel: Oh, I have a lot to say about this. Western medicine has historically been biased against behavioral medicine, stemming from Descartes' notion of dualism—the idea that the mind and body are separate. This view relegated mind-body interventions to the realm of psychologists and psychiatrists, while "real" scientists focused on the body. However, the brain and body have always been connected, and traditional approaches like meditation and mind-body interventions have clear effects on the body through neurohormonal pathways.

VR leverages these innate abilities, whether it's triggering the endocannabinoid system, endorphins, or changes in cortisol. It makes it easier for people to tap into these benefits without needing the extensive training of Buddhist monks, who average 30,000 to 40,000 hours of meditation to achieve similar neurological states.

With VR, especially when combined with biosensors like EEGs, we can monitor brain waves and achieve states associated with flow and calm. For example, the company Helium uses EEGs to reward users for achieving certain brain wave patterns, allowing them to control their VR environment with their minds. This integration of brain-computer interfaces and VR can unlock many therapeutic benefits.

Medicine finally recognized the value of these approaches after dismissing them for much of the 20th century. The pendulum is swinging back, and now therapies like VR are considered reasonable and mainstream, whereas 20 or 30 years ago, they might not have been as accepted. Things have indeed changed.

Praveen: I want to get back to GI. What advice do you have for private practice gastroenterologists? They're often busy with procedures like colonoscopies, so how could they incorporate something like this into their practice?

Dr. Spiegel: Absolutely. This is a perfect solution for the busy endoscopist. Many of us went into GI because we enjoy procedures and using our hands. We like stopping a bleeding peptic ulcer or screening for colon cancer. However, we also manage many patients with irritable bowel syndrome (IBS), which requires a more cognitive, non-procedural approach. This can create tension because managing IBS often feels like being an honorary psychologist, which isn't what many GI doctors signed up for.

The treatments for IBS, although effective for many, can be unpredictable. The relationship with the patient and their understanding of the treatment can significantly impact its effectiveness. Patients appreciate doctors who offer more than just pills; they value time and insights.

That's why we're developing an IBS VR program. It combines years of science on cognitive behavioral therapy (CBT) and mindful meditation into a program patients can use at home. This VR program allows patients to build skills like hypnotherapy, CBT, and mindful meditation focused on gut health. The GI doctor doesn't need to learn CBT or constantly refer patients to another specialist. They just need to introduce the VR program to their patients, who can then use it at home.

For private practitioners, this approach offers several benefits. It enhances patient care without adding to the doctor's workload. Patients are increasingly receptive to these tools and see positive results. Incorporating VR can improve patient satisfaction and outcomes, making it a valuable addition to any GI practice.

Praveen: The obvious follow-up question is, who pays for all this?

Dr. Spiegel: That's a great question. Currently, most insurance companies do not cover VR as a treatment, but some models are emerging. At Cedars-Sinai, we're launching a clinical VR program for both inpatients and outpatients, run by a psychiatrist trained in VR. He gets paid to deliver psychological treatments, which happen to use VR. Some psychiatrists are billing for VR exposure therapy for phobias, and there are codes being developed for VR in physical therapy.

The FDA is starting to look at these treatments, and Medicare may soon have to cover VR. AppliedVR, a company in LA, received breakthrough designation from the FDA for their chronic pain treatment program, which may lead to mandatory coverage by CMS. Other insurers, like Travelers Insurance and Blue Cross Blue Shield, are also considering support.

In the meantime, patients can purchase headsets for $200-$300 and download programs for free or for a small fee. So, while insurance coverage would be ideal, it isn't necessary for everyone to access VR treatments right now.

Praveen: How far are we, from a timeline standpoint, when all this goes mainstream? You might say it already is, but maybe if not mainstream, then I want your outlook on the future of VR.

Dr. Spiegel: It will depend on several factors. VR, in general, has become more mainstream, especially for gaming and entertainment. Two years ago, most people wouldn't know what VR meant, but now there's widespread familiarity. In healthcare, the catalysts for mainstream adoption will include insurance coverage and payment models, as well as more evidence around novel therapeutics and clinician acceptance.

This acceptance is already growing, and we'll see a gradual realization of VR's benefits. At some point, we'll reach a threshold where enough people know about and use VR routinely. In GI, particularly with IBS and disorders of brain-gut interaction, we'll see more effective therapies and patient demand will grow as they share their positive experiences on social media. Based on the rapid progress in the last two years, VR will penetrate healthcare significantly if this trajectory continues.

Praveen: This has been fantastic. I've been greatly inspired by your book. Any final words before we close?

Dr. Spiegel: No, I appreciate the time. If anyone's interested, the book is called "VRX: How Virtual Therapeutics Will Revolutionize Medicine." The book is really about what VR teaches us about our consciousness, the mind-body connection, and the boundaries of neuroscience. It explores the intersection between neuroscience, psychology, technology, and clinical medicine. It was a blast to write and is accessible to non-scientists interested in science in general. I hope you take a look and enjoy the book.

SECTION: VOICES V - ENTREPRENEURSHIP AND CAREERS

This section showcases the journeys of entrepreneurs and healthcare professionals who are driving change in gastroenterology, offering inspiration and practical insights for career development.

LEAVING PRIVATE PRACTICE TO PURSUE LIFE: DR. FEHMIDA CHIPTY

Dr. Fehmida Chipty is a Boston-based gastroenterologist, photographer, and facilitator. She founded Zamana Art as a creative platform for engagement, connection, and social impact.

Themes: *adaptation, physician versatility, physical and psychological stress of doctors, digital health, work-life balance*

Date: *May 2022*

Three Takeaways:

1. **The Importance of Adaptation in Healthcare:** Dr. Chipty emphasizes that the healthcare industry, particularly in GI, must adapt to technological advancements and changing patient dynamics. Without adaptation, physicians and practices risk falling behind as new methods and tools disrupt traditional approaches.
2. **The Value of Physician Versatility:** Physicians possess a wide range of skills beyond patient care, including business acumen, system analysis, and problem-solving. Dr. Chipty highlights the importance of recognizing and leveraging these skills in broader contexts, such as digital health and operational roles.
3. **Challenges and Psychological Stress in Medical Practice:** The conversation sheds light on the significant psychological stress and physical toll that high-volume medical procedures can have on physicians. Dr. Chipty advocates for open discussions about these challenges and the need for a balanced approach to work and personal life to ensure long-term well-being and satisfaction in the medical profession.

Praveen: Fehmida, let me ask you a question quite directly. Why did you leave the field of GI?

Dr. Fehmida Chipty: The 23 years I spent in my gastroenterology practice were wonderful. I had incredibly supportive colleagues and a group that was truly committed to high-quality patient care. However, there was a part of me that had been stirring to do something else for a long time. I had passions and interests that weren't being fulfilled. My journey into photography highlighted this for me.

People might argue that I could have done this on weekends while continuing my endoscopies during the week. And I did that for years. But I felt there was a part of me that needed more than what GI or medicine was giving me at that point. I wanted to travel and be there for my family; in many ways, they needed me now more than ever. Pieces of me were everywhere. Now, I realize those pieces are still everywhere, and that's just who I am. I am busy all the time, and I thrive on that. I needed some time and space to reflect on what these pieces all mean to me.

Praveen: But one would think that you'd be on perpetual vacation after leaving endoscopy. Is that not true?

Dr. Chipty: Leaving my practice was very hard. Making sure I left everything in a good place, especially for my patients, was very hard. My official last day was December 31, 2021. You might recall that on January 3, I called you and said, "Praveen, what can I do to help?" I have been very busy since. It took a couple of weeks to straighten out personal things because I had planned a trip to India in mid-January.

I've also gotten involved with a startup, where you're also involved. While working with the company, I thought, *This is an amazing organization that will impact many lives.* So, instinctually, after finishing my practice, I wanted to stay engaged and productive.

(Dr. Fehmida Chipty joined a wellness startup as a chief operating officer at that time and was involved in helping members reverse several chronic conditions)

Praveen: I want to take a step back to the field of GI and ask you openly what problems are on the ground. What are the issues that gastroenterologists and endoscopists face every day? Maybe they don't talk about it, but it's there.

Dr. Chipty: There are a lot of different issues, Praveen. These issues creep into your life slowly, and you keep saying, "It's okay, I'll manage, I'll do more cases, I'll see a few more patients." But it comes at a cost. When I first started my practice many years ago, I set up shop from scratch. There was no turnkey operation for me, and I let it build up slowly. It gave me a chance to figure everything out for myself, which I loved. I was able to run my practice with one person in my office and do just as good a job as in my last few years when I had a large operation in an endocenter with many people involved.

Not that everyone didn't have an important function, but there were layers of complexity added to the organization. The hospital relationships became more complex. Our hospital was acquired by a larger hospital system, and there was pressure to join the PHO. Our endocenter partnered with a larger management company, which made finances, relationships, and expectations more complicated. My practice and my partners did an amazing job balancing life and work obligations.

Praveen: When you say it became complicated, do you mean there was an unsaid push for more volume or more procedures?

Dr. Chipty: Yes, there's always an unsaid push for that. The relationship with the hospital also became more complicated during COVID. It was very stressful for physicians during the first few months of the pandemic. We had one physician in his 70s who worked just as hard, worried about his health, but showed up for his patients. We all did. I remember the fear my husband and I had about keeping our families safe—my elderly parents and his elderly mother. But we showed up, did our work, and adapted.

Physicians are always asked to adapt and think about what to do next to help the next person. That's our nature. But when will we say that we need something or need to reassess our lives? My mother's illness with COVID was a big factor, but this [change for me] was already coming. My desire to explore my creative side was also a big part of it. As a gastroenterologist, I hope all those procedures, lives touched, cancers prevented, and middle-of-the-night endoscopies made a difference in people's lives. But there were days I wondered if doing colonoscopy after colonoscopy was serving the greater good for me or others.

Praveen: Can you explain what you mean?

Dr. Chipty: If the end game for the bread-and-butter gastroenterologist is to prevent cancer, are there more effective ways to do that? That question is coming to the forefront now. Many of us do procedure after procedure and sometimes wonder if there are more efficient and effective ways to deliver population health, especially in terms of colon cancer prevention.

Praveen: It's not very polite to talk about money, but the advent of managed care ensured that physicians get compensated based on volume. From my observation, this seems to drive things. What's your view, or what did you observe on the inside?

Dr. Chipty: Absolutely. When people know that doing more procedures will earn them more money, that factors into their actions. Overhead costs keep rising, and reimbursements change. While we were always making a good living, the pressures to produce definitely increased. If doing more cases means earning more money, then the volume drive will happen. I don't think my colleagues were doing unnecessary procedures, but the volume is there, and it can be driven appropriately.

But let's talk about the physical aspect of this, which you might have had in mind. What toll does this take on our bodies? Beyond eleven procedures a day, it was physically hard for me. Most of my patients were women, and female colonoscopies are more complex. At one point, I started tracking how many of my patients with intra-abdominal surgeries. Many women have surgical abdomens, leading to more adhesions and technically more difficult procedures.

Before our unit was fully equipped with pediatric scopes, I used to get cramps in my hands after a day with ten colonoscopies. My hands hurt. As scopes improved and became more responsive to hand maneuvers, that got better. I had to learn ergonomics, be aware of my body position, and adjust my stretcher height. However, the neck position for the screen was a challenge if I shared a room with a taller colleague. I had to constantly think about all these aspects. If I didn't, I risked issues with my neck, back, and wrists, which is common among many in our field.

Praveen: Endoscopists talk openly about this. I know they've talked to me on the side. Why don't doctors talk about an issue that perhaps affects them and stresses them out every single day? I would assume it does.

Dr. Chipty: A few years ago, our PA left, and we had a very busy hospital service. We took turns being on service for a week at a time, and those weeks could be insanely busy. When I was on call at night, I would call the hospital doctor and ask, "How can I help?" They would never ask for help, and I never asked for help either. I would just say, "I can do this."

I don't know if we're afraid of looking weak, appearing incompetent, or if it speaks to our competence, but we don't ask for help. Physicians don't ask for help.

Praveen: When you look back and reflect on that phase, do you think it's a sign of incompetence for someone to ask for help or to say that the equipment or position doesn't suit their body? Or to question why we were doing so many procedures when it wasn't originally designed this way? What are your thoughts?

Dr. Chipty: Praveen. I think what you're implying is that this conversation needs to be had. Every endocenter needs to evaluate what they're doing and what they should be doing as a practice and as individuals. Someone who is physically bigger than I am may have more bandwidth to handle a certain procedure load. I'm not saying I'm able to do more technically than anyone else, but I remember a day when I had to cover for a male colleague. I had many more male patients that day, and it was physically much easier for me.

We need to consider what kind of patients each of us is getting. There are patterns. People seek out certain physicians, and all these factors play a role. We never make time to talk about this.

Praveen: I observe a lot of psychological stress among physicians. It's very apparent to me from the outside. But again, it seems to be because of the fear of coming across as not able or competent. Whatever the reason, it never gets talked about, and I see physicians suffering. I use the word suffering directly; they actually suffer a lonely battle.

Dr. Chipty: At the end of the day, it's about having that balance and being able to speak up for yourself. It's about having control over what you do day to day. If we can have a frank conversation about our personal capabilities—physical and emotional and how our work fits into our lives—and align that with what our practice needs, everyone will be happier.

Praveen: The industry is at a crossroads. Endoscopy physicians have dedicated 15, 20, 25, or 30 years of their lives to doing things a certain way. Now, there's artificial intelligence, digital biology with liquid biopsies, stool DNA tests, and more. The threat is looming. Scope Forward has always been about actively paving the way for the future, not being passive.

There are many threats: consolidation, health systems, big insurance companies, and a decline in patient trust. These factors must affect a physician's mindset about their field. It must take a toll on their mental makeup and on what medicine should be. Perhaps the entire industry needs to switch gears and think differently. I'm interested in your reflections on all of this.

Dr. Chipty: Most of us who went into GI love the hands-on aspect. We enjoy the procedures and the skill set we have acquired over the years. When I left, I was probably at the peak of my technical skills, which is bittersweet because I loved solving problems with my hands and mind. Many of us love the technical and results-oriented nature of procedures, so it's hard for gastroenterologists like me to leave that part.

Physicians are slow to change. It takes years for something in the literature to be implemented in practice. However, technology has been a major disruptor in a short time. If we don't adapt, we won't survive in our fields. The disruption will be imposed on us, like when a hospital is acquired or practices aggregate. Physicians tend to accept changes when they are imposed but are not proactive. That's where we need to change.

Praveen: There's a lot of self-doubt among physicians. They are some of the most brilliant people, yet there's a question mark about handling business or technology. You stumbled into the digital health world and are doing well. What are your reflections on this?

Dr. Chipty: We, as physicians, have accumulated a huge knowledge base regarding people, systems, and businesses.

It's become second nature. We wouldn't call ourselves businesspeople, but we are. We are psychologists, systems analysts, and IT people. We don't need more years of school to learn these skills; we've been living it. Assessing what we're good at beyond medicine is crucial.

Physicians can do far beyond just seeing patients or doing endoscopy. There's a breadth of knowledge we bring to any industry. The industry should realize that physicians with 25 or 30 years of practice are gold mines for solving problems and innovating. Any doctor is well-suited for business. Embrace the disruption, figure it out, and start the conversation. You might discover new abilities that take you to a new level.

Praveen: I was talking to a younger gastroenterologist who said physicians in his generation are just waiting to leave private practice. They find some parts of the work meaningless. My concern is if physicians leave the field, who takes care of patients? Any thoughts on this?

Dr. Chipty: It's very scary. We're all searching for meaning, and when we realize we have enough financially, we ask, what am I doing this for? The crisis in the world accelerates this thinking. I'm scared about where medicine is going. We need to take a hard look at our healthcare priorities and figure out how to meet the demand.

Younger physicians are starting with a work-life balance mindset, unlike our generation. They are also very savvy with digital tools. They might embrace digital disruption and develop population health solutions requiring fewer physicians. However, it will take a change from the ground up, including in medical schools and advanced practitioner programs.

If we don't address the provider-level question, this new generation won't stick around when pressures increase. They will pivot easily, unlike us. They won't put up with increasing demands at the cost of their work-life balance. If they leave, either we will find new solutions, or the healthcare infrastructure will crumble, and no one will be left to care for us. We need to address this now.

Praveen: This has been a fantastic conversation. Maybe there will be a part two in the future. I thoroughly enjoyed listening to your views. Thank you so much for sharing your views. Any final comments?

Dr. Chipty: Thank you so much for the opportunity. I always appreciate talking with you because I learn so much. I want to say to everyone (*referring to doctors such as herself*) that you have so much more to offer medicine and healthcare than you may realize.

BEYOND LINEAR HEALTHCARE MODELS: DR. DAN NEUMANN OF TRILLIUM HEALTH

Dr. Dan Neumann is the cofounder of Trillium Health. He's also the president and chief strategy officer at one of the largest gastroenterology groups in the Mid-Atlantic. Trillium Health focuses on custom diagnostics, R&D for biopharma partners, and personalized health testing for consumers.

Themes: *moving away from linear healthcare models, industry collaboration, direct-to-consumer lab tests, therapeutics vs diagnostics, biomarkers*

Date: *July 2023*

Three Takeaways:

1. **Innovation and Adaptation in Healthcare**: Dr. Dan Neumann emphasizes the importance of moving away from the traditional linear healthcare model to more innovative and adaptable approaches. He highlights the need for healthcare to keep pace with technology and social changes, stressing the role of physician-led innovation in improving access to care and wellness.
2. **Collaborative and Global Approach**: Dr. Neumann discusses the benefits of collaboration and data sharing in advancing healthcare. He envisions a future where healthcare systems are more interoperable, and physicians work with dedicated healthcare partners, rather than private equity, to ensure long-term, sustainable growth and better patient outcomes.
3. **Specialization and Procedural Focus in GI**: The future of gastroenterology, according to Dr. Neumann, will be procedurally oriented, with a focus on therapeutics rather

than diagnostics. He predicts increased specialization and regionalization within the field, along with a shift toward using physician extenders for initial and long-term care to allow gastroenterologists to focus on procedures.

Praveen: Dan, how did the concept for Trillium Health come about? What challenges in traditional models were you trying to address?

Dr. Dan Neumann: Well, you know, we still remain focused on delivering health care, and Trillium has really been something that came out of the disruption of COVID, understanding new ways to deliver care and wellness to people around the world. The traditional ways we're doing things in a linear model of delivering care aren't working, are limiting access, and are a very slow process. Trillium was an idea that we initially hoped to integrate into our traditional healthcare model. But as you know, traditional healthcare moves slowly, and the needs are really immediate. So, Trillium was born and has become a space for us to welcome the future of health and wellness.

Praveen: You were part of the GI mastermind, and it was fantastic to have you. Walk us through the path you took from that point in time that led to the founding of Trillium Health.

Dr. Neumann: That's so true, Praveen. This really came through the GI Mastermind, having that space to openly share ideas. These were things that either aren't talked about openly, aren't accepted openly in traditional healthcare, or are just batted around with a handful of forward-thinking leaders in the GI field and in medicine in general. Understanding different things we discussed in the GI Mastermind, from disruption of an industry to how to be part of a disruption rather than being disrupted. Our goal is to deliver aspects of healthcare and wellness to people—not just patients, but people. The model we're currently using is linear, and our ability to reach people where they are today is much different than before the pandemic or even a decade ago.

With technology and social media, the way people seek their care and wellness has changed. Healthcare hasn't kept up. My frustration with traditional medicine is that it's very slow to change. Physicians, in general, are also very averse to change. This was an opportunity to gather a few other forward-thinking physician leaders and rethink how we can meet people where they are. How can we, as physicians, step into the space where people are seeking out their care? If you're looking for health and wellness information, you should get it from trained professionals, not from the tech community or people without any training. There's a lot of misinformation being sold or peddled around the world that is not in people's best interests. It was time for physicians to take an active role in this disruption.

Trillium allowed a handful of us to begin that conversation. Now, it has grown into a space where we meet people where they are and deliver health and wellness in a non-traditional way.

Praveen: Just to set some context, when you say the old model of healthcare is linear, can you explain?

Dr. Neumann: Sure. If you think about what is involved in delivering care nowadays, it involves a person navigating the healthcare system, finding a provider, actually getting in touch with an office, and being able to make a reasonable appointment. In our society, and really around the world, there's an expectation that there should be an app that fixes everything or provides an immediate solution. But the traditional process takes time—getting in with a provider, coming into the office, taking time off work, sitting in a waiting room, and waiting for the provider. This can take an entire day. It's stressful, not accessible, and detrimental to people seeking care.

Then, the care that's delivered involves a provider sitting face-to-face with someone and setting up a plan for the next steps, which could be days, weeks, or months later. This one-on-one, linear type of relationship doesn't need to happen with the tools and technology we have available today. While that linear model still exists, we're going to see a shift toward non-traditional delivery options.

Praveen: Excellent. What does Trillium Health do today?

Dr. Neumann: It's interesting. We began Trillium as a platform to develop and identify biomarkers and bacteria in the microbiome, better understanding how the microbiome ties into disease processes. We hoped to leverage our understanding of disease processes to identify what's in the microbiome that might predict disease or new therapies. While the technology community has explored microbiome data through online kits or stool collection kits, it hasn't been deeply rooted in science.

For example, a patient might come in with a stool test listing 200 different organisms, and neither the patient nor we, as providers, know what to do with that information. It's more confusing than helpful. Trillium's initial goal was to take a serious look at the microbiome and other biomarkers in blood and tissues, aiming to predict disease and identify at-risk individuals based on their microbiome. We wanted to understand what a "normal" microbiome looks like in people without issues affecting their quality of life.

This way, we could target specific probiotics or bacteria to help patients reach that "normal" microbiome. Currently, it's hit or miss—patients are told to get a probiotic, but not all probiotics are created equal, and there's no regulation around them. People are being taken advantage of, and we see that.

Trillium began with this and other ideas but has evolved into a testing company that offers direct-to-consumer and business-to-business testing. We work on blood testing, innovative test design, and product development, including in the rare disease space. Interestingly, none of this is in the field of gastroenterology and has no overlap with the traditional medical practice.

Praveen: That's fascinating. I don't think you planned when you began that you would be creating a source of revenue beyond GI.

Dr. Neumann: We really didn't. We didn't start that way. But what drove that was the interest from companies wanting to partner with a laboratory that has physicians who understand the research, the science, and the importance of accuracy. They know the significance of quick turnaround times because time is crucial in diagnosing, treating, and predicting diseases. As physicians, we've been on the receiving end of these tests, so we understand the frustrations and the importance of reliable results. We can now collaborate effectively with industry and reach people around the world. This global reach will help us understand disease processes more comprehensively than we could within the confines of the United States alone.

Praveen: It's interesting that one of your first clients came from outside of the US. Can you talk about that?

Dr. Neumann: Yes, this came about through our relationships. Trillium met the needs of a large international company based in Spain that's been around for over a century. They were looking for a partner to help elevate their operations. They were trying to find people who might have a rare disease, and they needed to test a broad population. Traditionally, they were doing this through physicians and physician offices. Even though the test was free to patients and not insurance-based, they still had to navigate the traditional healthcare process to get the test.

We discussed with them our vision of bypassing these steps and going directly to people. By merging these efforts, we can conduct testing in a direct-to-consumer manner, providing results along with healthcare information to connect individuals to a healthcare system tailored to their needs.

This approach has dramatically expanded their reach. When searching for rare diseases, the key is reaching a large enough population. We're now able to help them by developing methods to deliver test kits to individuals, even in areas with limited access to traditional healthcare. We have the technology and partnerships in place to facilitate this testing in innovative ways.

Praveen: Let's talk about what frustrates you about gastroenterology as a space.

Dr. Neumann: Like I said, physicians are very slow to change. They get comfortable with what they're doing and practice for a long time. The traditional healthcare system in the US is also not easily adaptable, and it's been rapidly outpaced by technology. This is frustrating because, as providers, we need to recognize and keep up with the pace. For example, cancer screenings are moving away from invasive testing toward biomarkers in stool and blood. This will be the next phase of disease and cancer identification. However, gastroenterologists are slow to adapt, clinging to the belief that colonoscopy will remain the primary screening tool. This mindset is selfish and frustrating to me because our mission should be to offer the best care, not just stick to traditional methods.

Praveen: That takes me to a related question. What risks do you foresee for the field of GI for people who are not thinking the way you just described?

Dr. Neumann: There are several risks in GI right now. First, our workforce is not meeting the demand in this country, making it hard to find gastroenterologists. We're changing how we deliver care through alternative models and physician extenders, which is a good thing. The invasive procedures we currently use for screening need to change, and we should focus more on therapeutic procedures. If we don't adapt to these changes, traditional gastroenterologists will be left behind. There's also going to be consolidation in healthcare, and GI won't be spared from that. If we don't become part of these changes and technological growth, traditional gastroenterologists will die on the vine. We need to be part of the conversation and join groups of forward-thinking providers. Maintaining an independent physician practice is beneficial, but only if we are forward-thinking; otherwise, our independence will be taken from us.

Praveen: What's the worst-case scenario here for gastroenterologists?

Dr. Neumann: The worst-case scenario is that we don't pay attention, that we don't wake up and understand that collaboration and innovation are our friends, not our foes. There will be plenty of work for us to do, and we will do it more meaningfully if we can recognize this. But the risk is turning a blind eye or continuing to fight change and being left out of advancements.

Praveen: Someone out there may say, "Okay, Dan, I'm willing to listen to what you're saying. What's the best-case scenario here?"

Dr. Neumann: The best-case scenario is that we begin to share ideas and collaborate effectively. We stop viewing it as us versus them within practice models and start sharing data. We move toward more global, interoperable healthcare systems rather than sticking to proprietary ones. This shared data will make our work easier and better for patients. Ideally, physicians should look for partners dedicated to healthcare, not private equity. These partners should have a long-term commitment to healthcare. When we work with partners who truly understand and are dedicated to healthcare, our missions are aligned, and we can grow together sustainably.

Praveen: Where do you see Trillium Health going in three to five years?

Dr. Neumann: In the next three to five years, we plan to significantly expand the volume of testing and development projects we handle while providing very high-touch concierge solutions. We'll broaden our direct-to-consumer testing so that anyone around the world can access it. We will also conduct extensive research and scientific analysis on the data we collect, collaborating with our clients to enhance societal health, support population health initiatives, and help our corporate customers grow and innovate. These types of partnerships will drive the future of health.

Praveen: While the entire conversation has been about the future of GI, my last question usually is about the future of GI from your lens. What do you have to say?

Dr. Neumann: Well, the future of GI is still going to be a procedurally oriented field. It's going to be focused on therapeutics and not diagnostics. And we're going to see the consolidation of specialty services, like high-risk endoscopy and biliary services. We don't need to have those providers at every community hospital. We need to control the population and get people to specific gastroenterologists who specialize in those areas. So we'll see more specialization within our field and regionalization around that. We're going to see collaboration around shared risk models in reimbursement, and we'll need to have the data to show that we, together with payers and hospital systems, can work to keep healthcare costs down. We're also going to see a shift away from the linear model where gastroenterologists, trained as proceduralists, spend a lot of time in the clinic. We'll see more use of physician extenders for oversight, initial care, and long-term care, enabling us to get people cared for from a procedural standpoint faster. I also expect more population health collaborations with larger companies and industries, particularly in the key areas of disease prevention and screening.

Praveen: Great. Dan, anything else that you want to say in closing?

Dr. Neumann: I'll just say a huge thank you, Praveen, for getting these conversations out there. They'll spur a lot more discussion, which is always welcome. I'm always open to hearing from my colleagues around the world about ideas. If something resonates, bring it up because, as you've shown me, when you bring forward-thinking people together, great things happen, and it benefits our entire society.

Praveen: Excellent. Thank you so much, Dan. This was a lot of fun and very insightful.

Dr. Neumann: It's always great to be with you, Praveen. Thanks so much for all you do and for keeping the conversation going.

A DIGITAL STARTUP FOR WOMEN'S GUT ISSUES: DR. AJA MCCUTCHEN OF OLVI HEALTH AND UNITED DIGESTIVE

Dr. Aja McCutchen is a gastroenterologist, cofounder of OLVI Health, and executive board member of United Digestive. OLVI Health is a virtual digestive health platform providing personalized solutions for women.

Themes: *women's GI health, IBS, pelvic floor issues, reproductive health, digital health innovation*

Date: *April 2024*

Three Takeaways:

4. **Addressing Unmet Needs in Women's GI Health**: Dr. Aja McCutchen and her cofounders identified a significant gap in the understanding and treatment of women's gastrointestinal (GI) health issues. They recognized the unique challenges women face with GI symptoms and conditions, such as irritable bowel syndrome (IBS), autoimmune hepatitis, and pelvic floor issues, which are often under-researched and under-diagnosed.
5. **Leveraging Technology and Data for Personalized Care**: OLVI Health aims to bridge the gap in women's GI health by developing AI-driven pathways that utilize data analytics to provide personalized solutions. Their platform not only addresses digestive symptoms but also connects them to reproductive health, offering a comprehensive approach to women's health.
6. **Empowering Physician Founders in Healthcare Innovation**: Dr. McCutchen's journey demonstrates that physicians, despite their demanding schedules and

responsibilities, can successfully venture into entrepreneurship. By bootstrapping their startup and maintaining clinical practice, the founders of OLVI Health were able to retain control over their product and narrative while addressing a critical healthcare need.

Praveen: What's this new startup OLVI Health all about?

Dr. Aja McCutchen: I have about 40,000 women patient-years of experience. In the GI field, we've seen a significant demand not only for GIs but also for women GI specialists due to the unique needs women have. We face increased administrative burdens, long wait times—averaging about six months—and the complex mental demands of being a woman in GI. This journey has been both challenging and rewarding.

Over the past few years, various committees, summits, and groups have formed to address these unique needs, particularly for female patients. Along this journey, I connected with two other passionate women in GI: Dr. Asma Khapra and Dr. Latha Alaparthi. We officially came together as cofounders at the American Gastroenterological Association Women in GI regional conferences. Dr. Alaparthi is notable as the first woman president of DHPA, a lobbying and advocacy organization, and Dr. Khapra initiated a large PE-backed platform, a women's health network. I serve as the first woman on the board of directors for my platform company, which is the third largest in the US. Collectively, we share the passion and enthusiasm to make a significant impact, not just locally but on a broader scale. That shared goal has driven us forward.

Praveen: Awesome. And congratulations to all of you! Why are women's GI issues different, and who specifically are you targeting?

Dr. McCutchen: Certainly, the literature and data show us that women experience functional GI symptoms, such as irritable bowel syndrome, at a disproportionately higher rate than men. There are other GI diseases, such as autoimmune hepatitis, gallstone disease, and pelvic floor issues, that also disproportionately affect women. When we delved further into the challenges, we were appalled by the lack of data in this area.

Why are these changes occurring? Is it because women share the same compartment with reproductive organs? The data shows that about 46 percent of women overall experience GI issues, and 70 percent of women experience shifts in their GI symptoms during menstrual cycles. Yet, when you look for data to understand why this is happening, there's a paucity of information.

We started this journey into the Femtech space because we saw emerging initiatives, particularly out of Johns Hopkins, addressing why women have different presentations for cardiac disease. But we didn't see much in the GI space. When we looked at digital health applications through intense social listening with our women, we found that about 60 percent of them were using menstrual cycle apps, like the Flo app. There's a huge opportunity here. There's a big vacuum where GI symptoms that disproportionately affect women daily and go undiagnosed could potentially be addressed by technological advances.

Praveen: Awesome. What exactly are these solutions for women's gut issues?

Dr. McCutchen: Great question, Praveen. All of us in GI have seen the effects of misinformation being spread through social media and other channels where patients get their information. Part of our goal with OLVI is to drive people away from anecdotal evidence and move them toward evidence-based solutions. In our store, everything we offer is an evidence-based, validated solution based on rigorous scientific data. We don't want to just sell supplements, which can be dangerous as anyone can bottle up a supplement. We want to provide reliable, scientifically-backed solutions based on what a person is experiencing. That's the basic premise of the store.

Praveen: You've been a two-time participant in the GI mastermind, and we often talk about exponentials. How did you connect the dots, and how is this an exponential model?

Dr. McCutchen: Absolutely. We've all been trying to figure out how to go from serving one-to-one to serving one-to-many. As we built out our technology, we wanted to understand if people were more interested in community or being part of a cohort. We've created AI-driven pathways that use data analytics to come up with solutions. People can walk through the program sequentially, calculate their gut scores, and understand the contributing factors. This platform uniquely helps women understand the connection between their digestive symptoms and reproductive hormones, whether their cycles are contributing or if it's a perimenopausal effect. This will be a novel intervention, allowing women to go down different pathways using AI-based algorithms in a personalized manner.

Praveen: How did you figure out funding?

Dr. McCutchen: Every founder needs money. Every startup needs money. Initially, we did bootstrap. Bootstrapping is great because, as physician founders, we consider ourselves a powerhouse trio of female physicians who understand the market and its needs. We had to put on a business hat, learn to budget, and create our equity cap tables. I'm used to removing polyps and not dealing with cap tables, but understanding the fundamentals of business is essential when bootstrapping. I'm grateful we took this approach instead of doing an early partnership or receiving industry funding. Not that I'm against it, but putting our own dollars into what we believe in allowed us to maintain control over the product, narrative, story, and brand. We're excited to share the story behind the brand, and bootstrapping is a big part of that. So, for the majority of the early stages of development, we bootstrapped the project.

Praveen: I'm so excited to hear all this. You and the team have dispelled many healthcare myths. The number one myth is that physicians don't do startups because you're all so busy with your private practice world; everybody's head is in the endoscopy room. You've dispelled that myth. Despite your leadership roles, despite everything you do, you still have been able to figure this out. Number two is the myth regarding women in GI: that situations are altogether different for women with the responsibilities they have. You've dispelled that myth, too. I'd love to hear your reflections.

Dr. McCutchen: It's very interesting. Whenever you go to pursue something you're passionate about, you absolutely make the time. And this has come up, you know, over the past couple of months, as we are in our pre-seed round, you know, our capital raise is, you know, you have physicians that are still practicing. And as we talk about, Praveen, those closest to the problem are those closest to solving the problem. So, it's important for us to maintain those clinical touches so that we can continuously get feedback from the population that we're serving. And that is a strength in this particular platform.

We have three women with different practice environments. One of our cofounders is the CMO of a hospital system and has some flexibility in terms of administrative time and controlling the schedule. Being in private practice does allow some flexibility in the way we're scheduling now. Of course, we are still on an eat-what-you-kill model, but there is some flexibility in terms of predictably allocating hours toward the project. We've taken what would probably be a 70-hour workweek for a CEO and divided those tasks amongst ourselves. We've learned to organize ourselves as well. We use platforms like Monday for task organization. We tried different platforms, like Slack channels and Discord.

It's been such a steep learning curve, and it's incredibly exciting. And like I said, when there's passion behind it, you can stay up all night and do it. I don't think an entrepreneur in the startup phase is even interested in sleep.

Praveen: Aja, what advice do you have for people who might be thinking, *How can I be in her shoes?*

Dr. McCutchen: Well, you know, we always go back to the basic premise, Praveen, of understanding your WHY. My WHY has been that I've had a rich career making local and national impacts. But how can we continue to address the unmet needs of those we're serving? How can we ensure that everything we're trying to roll out, from an equity standpoint to access, is amplified and available to the masses?

To do this, you need to take the first step. My husband always says, "You can't get wet if you don't jump in the pool." So jump in the pool and start working on it. Put one foot in front of the other. There will be twists and turns, but the intellectual stimulation of learning how businesses work, how equity works, and how stock options work is empowering. It not only empowers you as a physician but also makes the bewildering aspects of practice management more comprehensible.

I encourage anyone passionate about a particular area or who sees an unmet need to go ahead and jump in the pool, get wet, and figure it out. One of the things I shared with my cofounders, a lesson from one of my mentors, is that entrepreneurs need to break things and break things fast. As physicians, we're used to taking calculated risks with a high risk-benefit ratio because we're dealing with human lives. This entrepreneurial journey allows for greater calculated risks, though we still need to be careful about data privacy and algorithm accuracy.

Building out great teams and bringing in other perspectives is crucial. Creating something from scratch that you're passionate about offers an unparalleled reward. It's exciting to see something tangible form out of conversations or initiatives like the GI Mastermind.

Praveen: What made the biggest impact for you from that program?

Dr. McCutchen: I encourage everyone to join the GI Mastermind. If you are a futurist, this is a group you should be in. We often operate in our local environments where we are the core decision-makers. However, to gain a broader understanding of the field's direction and potential disruptors and to exchange ideas with other brilliant minds, the GI Mastermind offers immense value.

The greatest value came from the intellectually stimulating cohort of individuals who not only talk the talk but actually walk the walk. Within that group, I saw people develop tangible products, services, and exponential technologies. I would discuss future innovations like a colon robot, and the next week, there would be a guest who had created a colon bot. It's like being part of a think tank that not only impacts the present but shapes the future of gastroenterology.

The experience in the GI Mastermind has been invaluable. I thank you, Praveen, and everyone involved in shaping this incredible program.

Praveen: My final question, Aja, is about OLVI Health. If you have to envision the future in the context of where GI is going and in the context of the future of GI, where do you see it landing? If you have to think freely, with no boundaries or limitations, three to five years from now, what's it going to be about?

Dr. McCutchen: That's a wonderful question. In the next three to five years, I believe we'll see emerging data that will finally help us understand the impact of socioeconomics on women's health. For example, we'll be able to understand what is happening with a woman's microbiome and how that impacts disease. This period will offer opportunities to close knowledge gaps that exist in this space.

For OLVI, the opportunities are limitless. Digital technology and consumer demands will drive what we provide, and since we are one of the first-to-market platforms, we are nimble and agile. Our strong social listening skills allow us to adapt and provide relevant services quickly. We have the support of our colleagues and can build teams efficiently. We also have the opportunity to create a multidisciplinary platform. We're starting in the GI space, but GI doesn't operate in a silo.

We want to link women's GI health with other factors, such as their reproductive system and how it might be tied to cardiovascular disease. The possibilities for OLVI are endless, and we can adapt to whatever model is relevant at that time.

Praveen: Excellent. I'm so excited that we had this conversation, and it's very satisfying to see OLVI shape up. We will be cheering you on.

Dr. McCutchen: Thank you so much, Praveen. Thanks for having me.

FROM THE ESOPHAGUS TO THE COLON, GI HAS VAST SCOPE FOR INNOVATION: DR. BARA EL KURDI

Dr. Bara El Kurdi is a gastroenterologist and host of the GI Startup Podcast. He's also a physician-scientist and innovator with experience in medical device development.

Themes: *digital health innovations, vast scope for innovation in GI, what young GIs want, ergonomics, GI startups, GI fellowship insights, endoscopy tools*

Date: *September 2023*

Three Takeaways:

1. **Vast Scope and Innovation in GI**: Dr. Bara El Kurdi highlights that the field of gastroenterology is incredibly vast, encompassing multiple organs such as the esophagus, stomach, small bowel, liver, pancreas, and colon. Each of these areas has its own unique innovations and advancements. This diversity ensures that there is a continuous stream of developments in GI, making it a specialty ripe for disruption and innovation.
2. **The Future of GI Beyond Procedures**: Dr. Bara El Kurdi envisions a future where the pressure on gastroenterologists to perform procedures is reduced, allowing them to focus more on patient care and overall community health. He suggests that changes in reimbursement models for colonoscopy could lead to a more balanced approach in the field. Technologies like AI-based tools, stool testing innovations, and ambient documentation can streamline workflows, reduce procedural burdens, and enhance the quality of care.

3. **Embracing Digital Health and AI**: The conversation emphasizes the significant role of digital health and artificial intelligence in the future of gastroenterology. Innovations in computer vision, natural language processing, and AI-driven diagnostic tools are set to transform how gastroenterologists diagnose and treat patients.

Praveen: Let's start with the GI Startup podcast. You've interviewed several innovators across gastroenterology. What did you learn from all these different conversations?

Dr. Bara El Kurdi: I've learned a lot. I could even say I learned as much as I did in my GI fellowship over the past three years from doing these GI Startup podcast interviews. The GI field is changing rapidly and is ripe for disruption. There's a vast array of innovations happening in GI, far beyond just endoscopy and endoscopic tools.

We have significant advancements in digital health, IBS, IBD, imaging, and devices. There's a huge variety of innovations, from bariatric endoscopy to third-space endoscopy. The tools used in these areas are very different, and there's a lot of innovation happening in all of them. Simple things like computer vision and AI-assisted polyp detection and characterization are also evolving. Now, we're even looking at using computer vision for dysplasia detection in Barrett's Esophagus and IBD.

When I started the podcast, I thought I might only have material for one year, releasing one episode a month. But three months in, I started getting so many emails from people interested in participating that I quickly had five years' worth of interviews lined up. It was eye-opening to see how many people are working on interesting things and the potential within the field.

I also learned that the GI innovation environment isn't limited to America or Europe. Amazing innovation is happening in New Zealand, Australia, India, Japan, and the Middle East. It's fascinating to see how vast the field is and how much innovation is taking place globally.

For young clinicians, especially those in training like fellows, it's essential to get involved and start looking around. The GI field is ripe for innovation. GI is heavily reliant on colonoscopy, and if CMS decides to reduce payments for colonoscopies, it will have massive implications for the entire GI ecosystem. Most of our income, both as gastroenterologists and as an industry, is built on this cornerstone.

We need to be prepared for potential changes and explore other areas that can be great sources of income and keep the ecosystem alive. Young clinicians should be aware of what's happening in the field. If your income is significantly reduced and you didn't see it coming, you only have yourself to blame. As specialists, we should be aware of these changes and have backup plans ready. It's our field, and we should know it better than anyone else.

Praveen: Awesome. Let's play a game, Bara. I want you to imagine a semi-circular curve, an aging curve. The beginning of the curve represents birth through adolescence; at the center is prime, and on the right side of the curve is aging and, eventually, death. Where's the average GI private practice space on this curve?

Dr. El Kurdi: That's a great question. I don't think there is much happening. We've reached a sweet spot, and unfortunately, we've become complacent. People know they can make a lot of money scoping every day, almost all week, maybe doing a half-day of clinic and then scoping the rest of the time. You can make a very good living this way. Because of that, people have gotten complacent and aren't branching out.

There's both a systemic problem and a tragedy of the commons situation. People are thinking, *I don't care what happens to everybody else; I'm just going to take advantage of this situation where I can do this many colonoscopies and make that much money.* The system isn't giving much incentive to innovate, especially for average small-scale private practices.

Things are changing. Our reimbursements are getting cut almost yearly, so people will start looking into other avenues. We need that push. But unfortunately, if reimbursement is cut abruptly, a lot of places will suffer catastrophic consequences. Hopefully, that won't happen. I do hope we'll be able to start integrating more innovation into small-scale practices. Digital health, in particular, offers a lot of room for this. There is hope.

Praveen: You made a career change recently. Why didn't you choose private practice GI? You chose a hybrid model so that you can innovate better. People out there want to know how you made the choices that you did, apart from family and personal reasons. What were the professional or industrial reasons for you to go down the path that you did?

Dr. El Kurdi: It's certainly appealing to go into private practice. The difference in the paycheck at the end of the month is a significant reason for many to choose that path. Private practice is essential—we have a shortage of gastroenterologists, and we need to screen many people. Private practice is noble and necessary.

For me, I wanted to contribute differently. I'm interested in innovating and creating devices, as well as developing digital health solutions, digital therapeutics, and AI products. I started thinking about whether private practice was the right environment for that. While many private practices participate in research, they often target profit alongside their research endeavors. This is essential, but I wanted an environment more focused on innovation.

I looked for places where I could practice clinical gastroenterology while interacting with biomedical engineers, electrical engineers, computer science majors, and software engineers. The best way to do that was through a hybrid practice. Virginia Tech Carilion Clinic was a great match for me. They have an excellent biomedical engineering program and a strong affiliation with a university program that supports these capabilities. This setup allows me to bridge the gap between clinical gastroenterology and innovation research.

Praveen: Well done. And congratulations to you for going down that path and having the courage, frankly, to do that because it's not easy. As a case in point, I meet many gastroenterologists who are quite advanced in their careers. And I mean, I hate to use the word stuck, but they are stuck in an existing model of GI care. They're so innovative as individuals. They have all these bunch of ideas, but they just can't wean themselves out of the endoscopy room. They don't have the time, and they're too exhausted by the end of the day. I can see all that innovation not coming to bear. And it is sad. It's actually quite sad for the industry more than anything else.

Dr. El Kurdi: I completely agree and understand that. That was always in the back of my head. I don't want to be one of those people who wakes up in 20 to 30 years and thinks, *Why did I do this? Why didn't I go after my ideas?* There's also the issue of a lack of training or education for clinicians, including gastroenterologists, on how to turn an idea into a product. Nobody really teaches us that, so there's a lack of belief that a good idea can actually become a reality. This is one of the biggest problems we face.

There are a handful of academic institutions that might provide some exposure to this process. I'm not aware of any structured programs that teach fellows or residents how to do this. However, in a few institutions, this happens often enough that people might get involved early in their careers or during their training. However, the vast majority of brilliant ideas from gastroenterologists never go anywhere because they don't believe their ideas matter or they're too exhausted and burnt out by the time they realize they do. They're stuck in the endoscopy room and don't have the time to make their ideas a reality.

Praveen: Very well articulated. A lot of people want to give advice to young GIs. They want to tell GI fellows what they need to do about their careers. Nobody wants to really sit and listen to young gastroenterologists and ask a reverse question: What exactly do you want? I want to ask that question. What do young gastroenterologists want from the industry?

Dr. El Kurdi: There are different kinds of young gastroenterologists. There's the kind who just wants to get as many colonoscopies done as they can before the system changes and retire very well, and that's totally fine. Then, there is a kind of young gastroenterologist who isn't really aware of what's happening with the entire GI ecosystem. They went into GI because it's one of the best and most lucrative specialties, but they're not really aware of the changes going on.

The third kind is the group this question really pertains to—the ones who are aware and want to do something about it but feel like they cannot. For these individuals, what's really needed from the industry and GI societies is to involve them more in what's happening within the GI space. It's tempting for a startup company to go to a key opinion leader with decades of experience because they have a lot of knowledge. However, the space is changing a lot, and many of these seasoned professionals don't have as much stake in the future of GI. They have very little skin in the game.

I'm not trying to criticize anyone. Almost every key opinion leader I know in GI is a great gastroenterologist who has made massive contributions. However, young GIs have a lot more skin in the game. They're going to be in this field for the next 30-40 years and have a vested interest in developing it. More involvement by these young physicians is crucial because they see a future that the older generation may not. The older generation has lived through a completely different GI environment and may not see themselves practicing much longer in this changing landscape.

Praveen: Very insightful comments, Bara. Can you present the landscape of AI in GI?

Dr. El Kurdi: There is a massive transformation happening in AI within the GI space. We're seeing an immense number of AI products, though many might not be particularly helpful in the long run. If we break it down, we can look at AI in terms of endoscopic and non-endoscopic applications, with non-endoscopic further divided into computer vision and natural language processing.

For endoscopic applications, which mainly involve computer vision, there is a lot happening in diagnosis—polyp detection and characterization, IBD activity assessment, dysplasia assessment in Barrett's Esophagus, and even recruiting patients for clinical trials based on the morphology of their disease. These applications will likely incorporate additional data, such as age, gender, comorbidities, and medications, to improve accuracy.

Recently, there have been discussions about addressing bias in training datasets. For example, smaller models could focus on specific populations to reduce bias. This is a promising area, but it's a bit far in the future due to the investment needed and the current lack of reimbursement structures for these technologies.

Outside of endoscopy, there are computer vision applications related to workflow optimization—like cleaning times, patient turnover in endoscopy suites, and imaging within the GI space, especially for IBD. Companies are working on MR enterographies and CT enterographies to find fistulas and early responses to therapy.

When we move to large language models and natural language processing, we see applications in diagnostics and workflow optimization. Diagnostic models can enhance diagnoses by analyzing vast amounts of data, including unstructured clinical notes. Decision support systems can collect data and suggest the best course of action, which the clinician can then confirm or modify.

Lastly, workflow optimization tools like ambient documentation can write notes during endoscopy procedures, and tools can help with prior authorization and managing in-basket tasks in EMRs. Overall, it's a vast field of innovation in AI.

Praveen: Superb response, Bara. I hope people realize that there are so many different GI startup ideas in what you just said. There's no dearth of revenues or innovation in the field. Let's get to your startup. I know it's in stealth mode, so please tell us what you can about what your startup does.

Dr. El Kurdi: The startup basically came out of my own frustration. As a young gastroenterologist, I want to be able to practice for as long as I can. When I started hearing about endoscopy-related musculoskeletal injuries, I got worried. The numbers show that anywhere between 50 percent to 75 percent of gastroenterologists end up with endoscopy-related musculoskeletal injuries. Looking at that as a young guy, I realized my chances were terrible. I don't want to suffer through this; I want a solution.

All the recommendations boiled down to this single picture we see in every ergonomics lecture—telling you to stand upright, set the bed height, and position the screen correctly. It felt like this wasn't enough. We'd get an ergonomics lecture once a year that lasted maybe 15 minutes, and for three days afterward, you might pay attention to your posture. But then old habits kick in, and you just do what you're used to doing.

I started thinking about how we could provide continuous monitoring and feedback with a complete analysis of what you're doing wrong and right in the endoscopy suite. We finally created a computer vision program that can analyze the posture of endoscopists as they perform procedures. It gives a comprehensive analysis of their posture, calculates the risk of injury using validated skeletal injury risk assessment tools, and provides a risk assessment for each joint. It also offers feedback on how to correct posture.

I partnered with a sports medicine physician and a rehab physician to develop an exercise program tailored for GIs based on injury prevention principles used in sports medicine. Athletes like basketball players face repetitive stress injuries, like what we experience in GI. Sports medicine has addressed these issues, but we hadn't adopted those principles. We used the same principles for gastroenterologists and incorporated them into the software.

Once this was built, I decided it was a good idea to create a company around it. However, I realized there are many other AI applications I want to work on and contribute to the GI space. Rather than building 15 startups, I decided to have a single startup focused on AI in GI, with multiple products. The ergonomics product is one part, and we'll have other parts working on workflow optimization and, hopefully, reimbursement optimization as well.

Praveen: I'm so glad, Bara, that you're bringing attention to this important area. My final question to you. Even though this entire conversation has been about the future, what is the future of gastroenterology from everything that you know, everything that you're seeing?

Dr. El Kurdi: There is a great future for GI. There's so much happening, and the field is so vast that it's extremely difficult for GI to have a bad future. Although colonoscopy and screening are big parts of what we do, there's a lot more to it than that. GI is a single specialty, but it encompasses so much.

If you look at other specialties in medicine, like cardiology, it's focused on one organ. But GI covers multiple organs: the esophagus, the stomach, the small bowel (which functions differently in various sections), the liver, the pancreas, and the colon. Each of these areas has its own innovations and advancements.

I'm hopeful that the future will involve less pressure to perform procedures. Changes in colonoscopy reimbursement could actually be beneficial, reducing the burden on gastroenterologists and opening up more opportunities for us to spend time with patients and contribute to community health. Technologies are being developed to make procedures like colonoscopy easier, whether through interventional innovations or AI-based tools for better lesion detection and polyp characterization. There are also advances in stool testing that could reduce the need for colonoscopies.

Other innovations, like ambient documentation, could free up our time from administrative tasks, allowing us to focus more on patient care. Digital health advancements will help us reach more people and provide better care to our communities. So, I see a future where GI is better for everyone—physicians, patients, and communities.

Praveen: Awesome. Dr. Bara El Kurdi, thank you so much for answering all these different questions directly and openly. It's going to be so helpful for people.

Dr. El Kurdi: Thank you, Praveen. Thank you for having me. This has been an awesome conversation. I loved it.

BUILDING AN END-TO-END VIRTUAL GI STARTUP: DR. RUSS ARJAL OF WOVENX HEALTH

Dr. Russ Arjal is the founder and chief medical officer of WovenX Health. He's a practicing gastroenterologist with interests in cognitive GI and digital health. WovenX Health integrates brick-and-mortar GI practices with a GI-trained virtual care team and technology platform.

Themes: *GI access, virtual care, hybrid model, white-labeled service for private practices, AI in GI*

Date: *August 2023*

Three Takeaways:

1. **Importance of Access and Efficiency in GI Care:** Dr. Russ Arjal emphasized the critical need to improve access to GI care, highlighting how delays can lead to severe health consequences. WovenX Health aims to address this by providing virtual consultations, which can significantly reduce wait times and expedite necessary diagnostic procedures, ultimately improving patient outcomes.
2. **Integration of Virtual and In-Person Care:** WovenX Health operates as a white-labeled service, partnering with existing GI practices to offer virtual visits. This hybrid model ensures that patients receive timely virtual consultations while still having access to necessary in-person procedures. This approach enhances capacity without compromising the quality of care.
3. **Future of GI with Technology and AI:** Dr. Arjal envisions a future where AI and advanced technologies play a crucial role in GI care. He predicts that GI practices

will become more cognitive and less procedural, with AI assisting in decision-making and protocolizing common GI issues. This shift will enable GI specialists to focus more on diagnostic accuracy and root cause analysis, aligning with patient expectations for comprehensive care.

Praveen: First of all, congratulations on the journey that you've undertaken of starting up a digital GI startup. It requires courage to let go of private practice GI and take the plunge. I want to understand this story a little bit more and get into your mindset. Was there a pivotal moment? Have you always wanted to start a startup?

Dr. Russ Arjal: It's a great question and something I reflect on often. It takes time to have clarity after making such a decision. I had the idea for a long time, starting around 2016, which was eight years after I finished my fellowship. By then, I had built up some experience and noticed the errors of access all around me while being focused on the rare diagnostic errors in GI. Innovative things were happening everywhere, but there wasn't much change in how patients accessed our care, leading to negative outcomes. Patients were metaphorically metastasizing in line, which was troubling.

Despite seeing advancements outside of GI focused on speed and efficiency, the pandemic was the turning point. It showed me that we could take care of most GI patients effectively without requiring physical exams. That realization gave me the courage to leave. It was a combination of ruminating on the idea and firsthand experience showing we could provide good care. In the end, it was something I felt I had to do, not a calculated decision. People thought I was crazy, and maybe they still do, but it's been a wonderful journey with no regrets.

Courage might be too strong a word; naivety also played a part. It has to be something you feel compelled to do, and you have to go all in. Building something from scratch can't be a hobby or a side gig. To create something transformational, you have to put all the chips on the table and commit fully.

Praveen: Can you share a little bit about what WovenX Health (formerly Telebelly Health) does today?

Dr. Arjal: Yeah, so we are about access. In general, we saw these market inefficiencies. I had seen them firsthand in GI for years, where it was incredibly hard to gain access to care, and that led to all sorts of problems. Our fundamental purpose, our mission, is to make access simpler and seamless.

When we launched, we had multiple choices for partnerships in the GI space: payers and other companies, but we chose to partner with practices. These practices often didn't have many partners and faced risks of disintermediation all around them. It's tough to be in an independent practice, but it doesn't mean you can't be successful; there's just a lot of friction.

We wanted to partner directly with practices to increase their capacity, and we thought we could achieve this with remote staffing. We source nationally for GI-experienced advanced practice providers and offer a white-labeled service, integrating them into local practices. This approach increases a practice's capacity to see patients and is dynamic—practices can scale up or down as needed. That's where we've started, and now we want to expand further, but these are our beginnings.

Praveen: To clarify, is this an entirely virtual program?

Dr. Arjal: It's a completely virtual program, but we have brick-and-mortar partnerships. We don't believe you can perform high-quality GI work entirely in a virtual space; you need those physical partnerships. We've partnered with innovative groups looking ahead and wanting to collaborate. We offer a white-labeled service, providing a turnkey operation where our APPs, who are based coast to coast, plug into local groups and see patients on their behalf.

Praveen: One of the biggest struggles in the GI space seems to be in terms of mindset. There are many talented and entrepreneurial GI doctors out there who want to do so much and are capable of doing so much, but money often gets in the way. Some think, *I could be making so much more doing endoscopy procedures, so why should I forego that revenue for a startup?* Others wonder where to generate the funds needed to pay salaries and hire people or how to go about fundraising. How did you solve the funding question for your startup?

Dr. Arjal: I asked for help. I listened and learned about the private equity world, which is different from venture capital. The psychology of venture is totally different. The macro-environment has changed a lot, and capital was more accessible for a long time. We started this at a time when it was harder to raise funds. I used my network and got a bit lucky. Some early investors were people who had started and sold companies and understood the venture space and digital health. I built a team and knew my limitations, so I gave up my CEO title early and hired Sheri Rudberg, who is very talented. Getting people who understand the venture space helps a lot. We leveraged our networks, found people who believed in our mission, and raised our initial tranche of angel funding. In the beginning, I simply asked, checked in with my network, met people, and got a little lucky. For instance, I met Trey Lauderdale, who founded and sold Voalte and is deeply connected. He helped me a lot at the start and mentored me. That's my story, but leveraging your network and the knowledge of people who know more than you is crucial.

Praveen: During that phase, were you playing Batman? Were you continuing your day job in private practice GI while building this on the side, or were you able to fund yourself to go all in?

Dr. Arjal: Yeah, I did self-fund at the very beginning, and then we were able to raise our first round. I still maintained a practice for a period of time. I still do clinical work, but it's a small percentage compared to my time at WovenX. Once I built a team and we had enough funding, I felt I needed to focus full-time. For eleven months, I didn't do a single colonoscopy. Some people might think that's crazy for a GI doctor, but there was too much work and too much to learn. So, for a brief period, I balanced both, but once I had momentum, I jumped in fully.

Praveen: You were part of the GI Mastermind cohort for 2022 and were an active participant. You always brought a contrarian view to our discussions. How did you use it to shape your thinking and WovenX (Telebelly)?

Dr. Arjal: The GI Mastermind course was very valuable for me. I joined at the perfect time. It offered a mindset very different from where I'd been for a long time. GI is a successful operation, but it's a narrow focus, primarily on endoscopic revenue. The Mastermind course introduced me to people thinking differently, taking chances, and having an exponential mindset. It reinforced that I wasn't crazy; there were others thinking the same way. It provided a broader worldview of things happening outside of medicine that will eventually apply to medicine. The community, exchange, and challenge were wonderful. People would push back on your ideas in a kind way, making you think. You need that as you move forward.

I believe you're remembered for the rules you break, as Phil Knight said. To do something transformational, you have to break the rules and be willing to fail. Most GI doctors wouldn't give up a comfortable life for something uncertain. The Mastermind course showed me there were others willing to take that risk. It was important and impactful for me to be in a space where people were thinking about doing the same sorts of things. It was wonderful.

Praveen: Very interesting to know. Let's get back to WovenX Health. You talk about access and solving the problem of access. I want to understand this more from the patient's point of view. I'm sure you saw the need and the trouble patients were having because they were not getting access to GI care at the right time. Maybe the wait was too long. Can you provide specific clinical examples?

Dr. Arjal: Sure. I'll start with my own experience. In 2016, I had a 42-year-old software engineer from Microsoft with rectal bleeding. Completely healthy, it took him three months to see me. His primary care doctor convinced him it might be a hemorrhoid, but it could be something else. Three months to see me, another three months to have a colonoscopy. During this seven-month wait, we discovered the source of the bleeding was rectal cancer. He had a young family, and his disease was metastatic at diagnosis. I can't say the seven-month wait changed his outcome, but it's a story I think about often.

Even in our first 500 patients at WovenX Health, we saw several malignancies. These patients' journeys went from four to six months to being diagnosed in several weeks. They saw someone a few days after the referral, and we scheduled procedures quickly. This change in timing potentially altered outcomes.

There's a lot of pathology in the queue waiting to get into a GI clinic. The wait to get into a GI clinic right now is anywhere from three to six months. During that time, terrible things can develop. We can see those patients much sooner, get them the care they need, and move them through the system. This benefits both groups and patients. Even replacing brick-and-mortar visits with GI APPs and moving that visit up by months can really change outcomes for a lot of patients.

Praveen: Let's say there's a patient who can't get an appointment. In the example you gave, what would happen in a WovenX world? How does the patient reach out to you, and what is the process from there? How does the APP interact with the patient?

Dr. Arjal: That's a good question. This is a white-labeled offering, meaning these are not our patients but those of the practice we partner with. The patient schedules through the practice, typically calling in for a clinic visit. They're then given a choice. Virtual visits aren't for everyone, and they aren't appropriate for every clinical situation. But often, patients just want to be seen sooner. So, they're given the option to wait for Dr. Smith, who might be available in 90 days in Lancaster, Pennsylvania, or they can have a virtual visit with one of our providers much sooner, often within days.

In this setup, the patient isn't told they're seeing WovenX; it's seamless, as we're an extension of the clinic. They get a virtual visit with our provider, who may be in a different time zone but is integrated into the clinic's operations. We use the clinic's scheduling system, document in their EMR, and provide full visibility for our partners. This allows us to see patients quicker, start their workup, and ensure they receive timely care.

Praveen: As far as labs and everything else, does the patient have to visit the practice for those, or is it all done remotely? How does that integrate into the practice's EHR in a virtual visit?

Dr. Arjal: Yes, we order everything through their EMR. So, if a practice uses Epic or gGastro, we order labs, imaging, or any necessary tests through their system. If we order a CAT scan or blood work, it goes through the labs or imaging centers that the practice normally uses. It's the same workflow—they just mirror it. For endoscopic studies, which are crucial for GI patients, we order them to be done at the ASC of our partner practice. Essentially, we are an extension of the practice, operating under their workflow and infrastructure. Patients see someone affiliated with the practice, which creates immediate trust. We have a strong focus on delivering high-quality care, and our quality program is robust to ensure we maintain A+ standards.

Praveen: What's the business model? How do you get paid?

Dr. Arjal: Initially, we thought we'd be a direct-to-consumer company, but the customer acquisition costs are very high, requiring significant capital. We quickly pivoted to a B2B model, working directly with GI practices. We assign billing rights to these practices and work within the insurance framework. Our providers are credentialed with the payers that the practices use, making them effectively part of the team. We manage the providers, handle cyber medical malpractice, and provide physician support. The practices bill on behalf of our providers, and we charge a per-visit fee. This model is simple and effective, offering margin benefits and significant downstream revenue for practices while allowing us to run an efficient operation. It's a win-win-win situation for everyone involved.

Praveen: As we wind down our conversation, Russ, I want to ask a broader industry-related question. What kind of problems frustrate you about the GI industry?

Dr. Arjal: One of my frustrations is that we're too comfortable. As gastroenterologists, we run the risk of being disintermediated because we focus so heavily on a singular test—colonoscopy. While we do many things in a GI practice, the economic model heavily relies on colonoscopy. There are non-invasive tests, such as DNA tests, RNA tests, and liquid biopsies, that could challenge our reliance on colonoscopy. I don't think colonoscopy will be disrupted entirely, but there will be pressure on it.

Physicians, including myself, often resist change. This resistance can result in losing control and agency as others adapt faster. The GI space will evolve significantly in the next three to five years, and we need to be adaptable and open-minded to keep up.

Lastly, I often heard in board meetings at my former practice that physicians would say, "That's not what I want to do." While the market may tolerate that for now, it won't last forever. We need to be willing to change our thinking and approach. If we don't, we risk becoming less important, losing our seat at the table, and possibly some of the financial benefits of being in this space. It's been a long bull market for GI, and there's still room to grow, but we must adapt.

Praveen: What you said about the meetings, where someone might say, "But that's not what I want to do." Can you explain that a little bit more?

Dr. Arjal: Sure, so there are things that patients want that physicians don't necessarily want to provide. A great example is non-invasive stool testing, like Cologuard. About 95 percent of patients aged 45 and up have seen an ad for Cologuard, and its market share is increasing. Many GI doctors have pushed back against Cologuard, arguing that it is not the same as a colonoscopy. And that's true—colonoscopy is the gold standard. GI doctors often believe that everyone should have a colonoscopy. While that's a nice idea, it's a fantasy that will never happen. There aren't enough GI doctors or endoscopy spaces to perform all the necessary colonoscopies.

So, if I suggest at a board meeting that the best way to screen a community is a two-step process, offering patients a choice and performing colonoscopies for those who need and want them, I would get a lot of pushback. The argument would be,

"That's not what I want to do," because colonoscopy is considered the best test. However, the reality is that what patients want is more important and will drive the market. If patients are not going to get screened without an alternative, offering them a test like Cologuard that they will accept is crucial. The best screening test is the one that gets done appropriately.

So, if we focus too much on what we want as clinicians and ignore what patients want, we risk being outrun by other competitors who prioritize patient needs.

Praveen: In the GI Mastermind program, we talk a lot about mindset. So, I want to touch upon that. Why do you think physicians, especially gastroenterologists, are hesitant or afraid to embrace new approaches?

Dr. Arjal: There are a few reasons. One major factor is our training. Getting into GI is very competitive and has been for the last 15 to 20 years. There's a lot of talent in GI, and younger doctors have worked hard to get here. It's not easy—GI has an older median age of about 57 and a half, which doesn't help patients either. It's a lot of training that leads to a comfortable existence, but it also relies on colonoscopy maintaining its current status.

Changing paths as a physician is difficult after spending so many years becoming an expert. It's an apprenticeship, and you can't get those years back. Another part of it is the personality type of doctors. Many who enter medicine might have chosen finance or another field if they were more inclined to take risks. They tend to have a more cautious mindset.

So, part of it is the extensive training and the investment they've made in their careers. Another part is that physicians, in general, tend to be risk-averse. They are careful and meticulous, which is necessary in medicine but makes them hesitant to embrace significant changes.

Praveen: The training teaches physicians to be so for good reason. So, talking about the industry risks that you touched upon, how many years do we have before you sense disruption at scale?

Dr. Arjal: It's hard to predict exactly, but we'll see significant changes in the next three to five years. Colonoscopy won't be completely disrupted—there will always be a role for it. However, average-risk patients may increasingly choose liquid biopsy and stool DNA or RNA testing. These non-invasive tests will likely gain more market share over the next few years.

We'll still be performing many diagnostic colonoscopies, as there are still around 45 to 60 million Americans who haven't been screened and need to be. However, I believe the market share for average-risk screening colonoscopies will change. Liquid biopsy presents a potential existential threat, and Cologuard 2.0, assuming it gets FDA approval, will continue to capture market share. Patients prefer non-invasive testing, and that trend will continue.

If I were a young physician entering GI, I would be thinking about the next five years. The field will look very different. Private equity groups have a shorter horizon, maybe five years, while venture groups might have an eight to ten-year horizon. Young physicians, on the other hand, have a 25-year horizon. This mismatch in timeframes affects how we should prepare. In five years, I expect GI will involve more diagnostic studies and possibly become a more cognitive field than it is now.

Praveen: You touched upon a lot of these themes, but I wanted you to summarize and say it. If you have to reimagine GI care for the future, what is the future of GI?

Dr. Arjal: Reimagining care for me is still centered on access. You should be able to get a GI clinician on your terms within minutes. That's something that will happen eventually and is possible. Those visits will be smarter, with structured intake processes. You can protocolize most GI visits, which involve structured problems. AI will be crucial. I think we'll all have an Iron Woman or Iron Man suit on as we make decisions in five years. This will happen not just in GI but in other verticals as well.

All the principles we're talking about in GI apply to most other specialties, too. We won't be the first wraparound practice company out there; there will be others in multiple specialties. GI, in the bigger picture, will become a more cognitive field. Right now, we've been focused on a singular test, which is very useful and life-saving. However, there will be different screening programs and more two-step screenings. We'll see more value in the space, pushing us to take on more diagnostic studies.

I think we'll pivot back to the roots of GI, which, for the older generation of gastroenterologists, was much more cognitive and less procedural. We'll still do procedures, but we'll be more focused on the clinic space, thinking more about root cause analysis because that's what patients want. We have a procedure mindset right now, and while procedures are important and helpful, patients often have a root cause mindset. Those patients aren't well served by practices or clinicians. I don't think I did a great job in my career answering those questions. They're hard to answer, so I think we're not going to pivot—we're going to morph.

Praveen: Awesome. As with every interview I do, I end up learning a lot. Your perspective was very clear and open, and thanks for sharing everything you did.

Dr. Arjal: Praveen, pleasure. Thanks for hosting me, and thanks for starting all these conversations. The GI Mastermind has been great, and I appreciate you allowing these sorts of conversations and giving us a platform to talk and build a community. It's really helpful.

Update from Dr. Russ Arjal (June 2024)

We are on track to achieve over 50,000 GI patient encounters (patient visits and smart digitized intakes) in 2024, a huge leap from fewer than 200 in January 2023. This rapid growth highlights the effectiveness of our integrated model, which was developed in partnership with innovative specialty groups to benefit all stakeholders. With signed contracts in place, we are poised for significant expansion in 2025, further solidifying our leadership position in the field.

Since our last update, we have launched On-demand GI care, reducing the median time to consult to minutes rather than months. Patients love the emphasis on speed, reflected in our net promoter score (NPS) score of 90+, and practices appreciate the return on investment. The cost savings for payers are also significant due to a reduction in ER utilization. We plan to publish data on this later this year.

Additionally, we now have an analytics platform and will soon roll out our AI suite. We are excited about the future of our platform in GI. We also plan to enter into other specialties this year.

SECTION: VOICES VI - PERSPECTIVES

This section provides a broader perspective on the future of gastroenterology and healthcare from leading voices in the industry, reflecting on the challenges and opportunities ahead.

"GI FACTORIES" ARE MORE VULNERABLE NOW: DR. LAWRENCE KOSINSKI OF SONARMD

Dr. Lawrence Kosinski (Dr. K) is a gastroenterologist and founder of SonarMD and VOCNomics. SonarMD is a value-based care coordination program for digestive health, partnering with health plans and providers to deliver proactive, cost-effective care. VOCnomics provides real-time analysis of volatile organic compounds in stool to offer actionable health insights.

Themes: *value-based care, digital health, innovation in GI, private equity, mastery and career management for young and mid-career professionals*

Date: *September 2022*

Three Takeaways:

1. **Embrace Innovation and Mastery:** Dr. K emphasizes the importance of mastering one's field while also being open to innovation. Mastery provides the foundation for confidence and efficiency, which in turn creates the bandwidth for exploring new initiatives and technological advancements. This balance is crucial for personal growth and advancing the field of gastroenterology.
2. **Adaptability in a Changing Landscape:** The interview highlights the significant shift toward digital health and the impact of technological advancements on traditional practices. Dr. K points out the vulnerabilities of GI practices that remain focused on traditional procedures and stresses the need for adaptability. Embracing new technologies, such as less invasive screening tests and AI, is essential for staying relevant and providing better patient care.

3. **Strategic Career Management:** For mid-career and younger gastroenterologists, strategic career management is key. Dr. K advises focusing on areas of genuine interest, learning to say no to avoid overcommitment, and planning for the future. This approach allows for professional growth, work-life balance, and the ability to innovate and contribute to the field in meaningful ways.

Praveen: After a long and successful career as a gastroenterologist that led to a national leadership position, you've turned your interest to value-based care. In 2014, you founded SonarMD, which develops the next generation of chronic care management for patients with high-beta chronic disease. Today, Dr. K, you're going to talk about yet another startup—a smelly one at that—and I'm excited to learn more. Can you give us an update first on Sonar?

Dr. Lawrence Kosinski: Thanks for asking, Praveen. Sonar continues to grow. I spend most of my professional time speaking to health plans, and we've been bringing health plans into the Sonar funnel and expanding further. We've been operational in New Jersey, Minnesota, Illinois, and California. In Illinois, we've revamped our original contract into a new one, and we have many more in our pipeline that are very close to being operationalized.

I'm still involved in national leadership; I'm back on the AGA governing board as the board counselor for development and growth. I also sit on the Physician-Focused Payment Model Technical Advisory Committee (PTAC), which reports to the Secretary of HHS. So I'm still involved nationally, Sonar continues to grow, and I'm having fun, but I have other initiatives that are growing as well.

Praveen: How do you manage to do all this and juggle so many roles?

Dr. K: I live by the "one wife, one child" concept, so I guess that gave me a lot of extra time. But you know, there's a lot of time in everyone's life, and I try to be as productive as I can with the time I'm allotted. I'm constantly thinking of new things to get involved with, so I find the time to do these things.

Praveen: You've referred to private GI practices as "colonoscopy factories" earlier. This is always an interesting topic, considering the industry's digital shift. Do you think most gastroenterologists are aware of these changes, or are they too busy with endoscopies to notice?

Dr. K: I don't think anything has changed in the GI practice world. COVID was a temporary assault on GI practices since much of their income comes from elective procedures, which went away during the pandemic. Now, GI practices are extremely busy catching up on colonoscopies for people who put them off during COVID-19. Plus, the screening age has been lowered to 45, so there's been no reduction in the focus on colonoscopy. GI practices are still "colonoscopy factories," if not more so than before.

As for digital, what you're seeing is a digital world developing outside of GI practices. Companies in the GI space are disintermediating GI practices by setting up direct-to-consumer (B2C) businesses. They are going around GI practices. Unless GI practices take their heads out of the sand, there will be other arrangements. Large self-funded employers are becoming providers—"payviders" employing providers in their retail outlets and building on primary care. They won't knock on the doors of individual GI practices to participate in their national networks; they'll engage large companies industrializing this space while disintermediating GI practices.

This is a big threat to GI. Companies are starting in the functional disease space, like IBS, and nibbling at the IBD space. This isolates GI practices, making them more like colonoscopy factories. Patients want services when they want them and want to pay as little as possible. The proprietary nature of private practices makes it difficult for patients to satisfy those needs, so they turn to B2C companies. These companies engage patients, but I don't know if that's a sustainable business model. Time will tell. Ultimately, they may need to be married to the provider space as an adjunct, but right now, GI practices aren't engaging much with the digital health world. They're doing colonoscopies because that's where the money is, and that's what's driving activities.

Praveen: Have GI practices become more vulnerable or less vulnerable in the last couple of years?

Dr. K: They're more vulnerable. They are still very narrowly focused on a single revenue stream that is elective and vulnerable to technological advances. You're seeing liquid biopsies coming into the market now. Non-GIs and even non-physicians are starting to perform colonoscopies. The market will find the lowest-cost way of providing services. Companies know that patients will prefer a blood test over a colonoscopy.

Colonoscopy will remain a strong revenue stream for the next few years, but there are vulnerabilities on the horizon. The infusion business is already vulnerable. Trends in biologics and small molecule drugs show a shift toward self-administered and oral medications, causing IV infusion centers to see declining portions of the biologic space. That revenue stream is vulnerable. The functional bowel disease space is moving toward B2C with third-party companies. So, the GI space is more vulnerable today than it was before.

Praveen: Do you think this disruption, though inevitable, will appear as a sudden, unexpected change?

Dr. K: Yes, a technological advance can make it look sudden. For example, CMS has already approved the use of blood-based screening for secondary levels. Once one of these methods catches on, you'll see a significant drop in traditional procedures, and everyone will wonder where it came from. It will seem like it happened suddenly.

Praveen: Very interesting. Startups in stool RNA testing, AI in GI, and other areas are leading to further advancements. How do you think all this will play out? Can you share more about the development of your new startup, VOCNomics?

Dr. K: There will be a steady trend toward less invasive screening tests and more sophisticated blood-based tests. The industry will move in that direction. That's one of the reasons I got into the stool space and volatile organic compounds (VOCs). At Sonar, we're detecting earlier based on symptoms, but I wanted to see if we could identify patients even before they become symptomatic.

During COVID, I explored how to detect changes in stool smell using air quality sensors. Nurses and GI doctors can often identify conditions like GI bleeds or C. diff by smell alone. Patients also report changes in stool smell during IBD flares. I thought that maybe these sensors could correlate with VOCs released in stool from fermentation by the fecal microbiome.

I experimented with six air quality sensors in my bathroom, monitoring my fiber intake and the VOC readings over six weeks. Two of the sensors correlated nicely with fiber intake. After consulting with experts and creating a prototype with a bioengineer friend, we developed a device that measures VOC levels in the air during bowel movements. The device sends readings to the cloud for data analysis using AI.

We are now conducting multiple studies to validate and replicate my initial findings. I filed a provisional patent and started a company, VOCnomics, to hold the intellectual property. This initiative helps move earlier in the disease process and has potential applications in various conditions, including IBS, diverticular disease, colon polyps, Parkinson's disease, diabetes, and cardiovascular disease. It's exciting to have so many interesting things to work on, even at my age.

Praveen: What an amazing story. I want to ask a question on behalf of many younger gastroenterologists out there. Unlike the earlier set of gastroenterologists who stuck it out in private practice, this younger group is alert to what's going on digitally. They want to innovate and do what you're doing, but not everyone can. They struggle with time, loans, productivity targets, and day-to-day issues. What would you advise someone like that?

Dr. K: I'll use myself as an example. I went through the same thing. You finish your training, you're 32 years old, you have loans, maybe a young family, kids, obligations, mortgages, and you need to put money away for college funds. Maybe you have your parents to worry about, too. You're sandwiched between the kids and the parents. I totally get it.

The best advice I can give to young GIs is to master what you're doing. You invested a lot to get where you are. You studied and worked hard. Master what you're doing. If you're doing 15 colonoscopies a day, get really good at it. Hone your skills to do it efficiently; that will buy you the time you need.

Stay reasonable with your finances. Don't overextend yourself with huge mortgages and payments that don't leave you time for other interests. Create some time for yourself to follow your initiatives. Keep your eyes open to things happening around you. For example, I stumbled upon VOC monitors while reading an ad.

Give yourself the time for freedom of thought so that you do not miss opportunities. Go to meetings and talk to people. You'll learn from everyone around you. That's my best advice.

Praveen: Let's talk about the group in their 50s—mid-career GIs. What advice would you have for mid-career gastroenterologists who are so busy that they don't have time to innovate?

Dr. K: You do not have to feel like you have to do everything. Pick things that really interest you and focus on them. You have to learn to say no. That's been one of my biggest problems—I say yes all the time. My wife has taught me to say no. Recognize your limitations and don't suffer from fear of missing out (FOMO).

This is especially relevant for GIs in their 50s. They can see the end, even if it's far off, and need a game plan to get there. I set up a plan and stuck to it. When I turned 60, I stopped taking calls, going to the hospital, and taking new patients, even though it meant less income. This allowed me the bandwidth to pursue other directions. I wouldn't have Sonar today if I hadn't done that.

Over the next few years, I gradually cut down even further and focused more on my other initiatives. I planned to retire in 2018 but stayed another year to help my group negotiate a private equity deal. I was ready to retire at 67 but stayed until 68. I was only part-time throughout my 60s, which wasn't taxing and allowed me time and energy for other things.

Praveen: Let's shift gears to private equity (PE). How has PE played out?

Dr. K: It's done what we expected it to do. If someone puts money out for you, you're going to go after it. It's only natural that there's been growth in private equity driven by the financial incentives to sign on. There are about 7,000 private practice gastroenterologists in the country, and maybe 1,500 of them are in a private equity-owned practice. The great majority of GIs are still not in that, but we've seen growth and exactly what we anticipated.

Now, we're seeing the maturation of the early practices with a couple of second bites that have occurred. We'll have to see how this second round of investment in these private equity-owned practices materializes. Do they continue growing and adding practices, or do they transform and build themselves on a more quality basis to occupy an earlier position in the value chain? I hope it's the latter, but I've yet to see that.

Praveen: Do you think the private equity game will continue for the next several years? If so, how long? Or do you think we'll see other alternative models?

Dr. K: The way the first models rolled out won't be the model in the end. You're already seeing hybrids. You're going to see venture capital enter the space with more of a long-term, higher-return investment model. Companies are already doing this in non-GI spaces, but it will enter the GI space as well. If that can be used to build the clinical infrastructure so that the true value of the entity is more than just the sum of its parts, then that's worthwhile.

Investment phases will continue, and practices will keep consolidating, but you'll see different models emerging over time.

Praveen: I have a fundamental question. When I talk to physicians, there's a lot of conditioning at play that's been ingrained over decades. This conditioning limits their mindset to the existing norm and model, even though they see the emergence of digital health and new startups doing things differently. Despite seeing these examples, physicians seem limited by boundaries set by the industry or colleagues. Why is that? What is the underlying fear here that keeps the physician mindset so narrow?

Dr. K: Yeah, you're correct that this exists. You have to realize what type of person becomes a doctor and what they have to do to get through the process. They have to develop habits that help them do well in school and remember what they need to know. Doctors get very accustomed to certain drugs they use. Just ask a pharmaceutical company how difficult it is to get a doctor to change what they're doing because we want to feel comfortable with what we're prescribing. We build a comfort zone around it.

Physicians are besieged by the need to master a vast amount of knowledge, and they develop ingrained habits to succeed in their profession. When asked to step out of that comfort zone, there's pushback because it's uncomfortable. Your observation is totally accurate. The type of person who goes through education until their early 30s and then goes into practice and succeeds has learned to control risk and exposure. They continue to build ingrained habits, and it's hard to get people to change out of that. That's why I say master something, even if it's uncomfortable.

Praveen: That would have been okay for the last 15, 20, or 30 years, but now, with the shift to digital, it's concerning. These changes are going to be sudden, and they could have a psychological effect on physicians. A new technology could tell them that all their mastery and hard work don't matter and they need to unlearn everything.

Dr. K: I've been very opposed to companies disintermediating doctors. At Sonar, we work to be an extension of doctors. Companies that figure out how to help doctors expand their technological space will succeed the most. The physician community is highly intelligent and capable of expanding into the technological space. Technology should gradually be integrated into what they're doing rather than forced upon them. AI, for example, will assist humans in their work, making us better rather than replacing us. Companies must understand that their technology should be part of the entire healthcare solution. We're going to go through some ugly times before we get there.

Praveen: Dr. Kosinski, thank you so much for sharing your thoughts and insights. Any final words of wisdom as we close this interview?

Dr. K: Have fun. Innovating is fun. Finding solutions is fun. Don't be afraid to try something new. People may think you're crazy, but if you're right, you'll have a good time.

THE FUTURE OF GI REMAINS BRIGHT: DR. SCOTT KETOVER OF MNGI DIGESTIVE HEALTH

Dr. Scott Ketover is a gastroenterologist and the president and CEO of MNGI Digestive Health, one of the largest and most respected independent practices in the United States.

Themes: *telehealth, independent private practices, data, comprehensive GI care, collaboration, expanding the scope of care*

Date: *April 2021*

Three Takeaways:

1. **Vulnerability and Resilience in GI Practices**: Dr. Scott Ketover highlighted how the COVID-19 pandemic exposed the vulnerabilities of GI practices, especially those in independent settings. The drastic reduction in ambulatory endoscopy during lockdowns underscored the need for practices to reassess their revenue models and prepare for future disruptions.
2. **Data Integration and Digital Health**: He emphasized the critical role of data in the future of GI care. He advocated for the integration of data across practices, regardless of their organizational structure, to enhance clinical outcomes and cost-effectiveness.
3. **Expanding the Scope of GI Care**: The discussion stressed the need for gastroenterologists to look beyond traditional procedures like endoscopy and consider the entire GI tract, from mouth to anus. He mentioned the importance of developing comprehensive care programs for conditions such as advanced esophageal disease, liver disease, and functional disorders.

Praveen: Since the last time I interviewed you for my book, *Scope Forward*, the GI landscape has changed dramatically. What are your thoughts on these changes?

Dr. Scott Ketover: No one could have predicted COVID-19, a pandemic, or the lockdowns and their impacts since we talked pre-COVID. The pandemic has had a dramatic impact in a few areas. For gastroenterologists, especially those in independent practice, it proved how vulnerable we are in terms of revenue. A year ago, at the height of the lockdowns, there was essentially very little ambulatory endoscopy happening, which is, from a financial viewpoint, the lifeblood of today's GI practices. This situation was a quick awakening to the fact that revenue could drop significantly, directly affecting the personal incomes of physicians in the independent world. Physicians in employed settings were also challenged because their institutions were under stress.

COVID-19 was a rude awakening, but it also caused us to step back and reassess our vulnerabilities, which we had only discussed theoretically before. If we can't see patients face to face, what do we do? How do we provide care, take care of our patients, support our employees, and look after ourselves?

Praveen: One of the key reasons why practices or businesses consolidate in healthcare or private practice is the leverage gained with insurance companies and local hospitals. Given that you chose to remain independent, how have you managed negotiations with insurers and health systems in your region?

Dr. Ketover: Yeah, good point. I've come to view that type of leverage as negative leverage. For provider groups, the leverage usually means saying, "We won't join that network," or "We'll leave that network or hospital system unless we can negotiate a favorable contract." It's a binary decision; you're either in the network or not. If you leave, you don't enhance the care within the systems you're negotiating with. However, it's been the main lever that independent practices use when negotiating with payers.

Even with our size—85 gastroenterologists, 900 employees, and nine locations—we're dwarfed by the payers, hospital systems, and even the PE-backed MSOs with 300, 500, or 1,000 physicians. While having a large number of physicians provides some leverage in staying in or leaving a network, it doesn't necessarily translate to positive outcomes. The future lies in making our leverage positive, meaning we bring more value to our relationships and steer the direction of value-based care rather than just reacting to it.

Praveen: Do you think our dependence on insurance systems will decrease over time, stay the same, or increase?

Dr. Ketover: GI still largely remains a referral-based practice in most areas. Gastroenterologists depend on primary care referrals for patients. We have a strategy to increase our self-referred or independent patients and families, but I believe insurance and third-party payers will continue to play a significant role. There's some talk about expanding Medicare down to age 60 or potentially age 55. This would certainly increase the percentage of government-paid patients we see and change the payer mix. However, I don't think we're going to eliminate commercial insurance companies in the near future, and certainly not in my career.

Praveen: Considering digital health, over $20 billion was raised during the COVID period in 2020, with over 600 deals. These companies seem to be serving the same patients or consumers that GI and other specialties are serving but through a different model. They seem to be finding new business models and approaches, while GI and other practices appear to be sticking to their traditional methods. Am I thinking correctly, or am I completely wrong?

Dr. Ketover: No, you're on the right track. We've been fortunate for two decades that endoscopy procedures have driven the revenue side of independent GI practice. These procedures need to be done face-to-face or in person. However, there are threats to that volume. As technology improves and new screening methods develop, there will likely be more ways to stratify risk for patients and families, which could negatively impact the volume of screening colonoscopies.

I often discuss with my partners that if we look at our compensation for professional services, it's less than half of our total compensation. From a financial perspective, the work of a gastroenterologist is supported by the ancillaries generated by professional services. This includes procedures, pathology, anesthesia, radiology, pharmacy, and infusions. It's almost like a pyramid. Moving forward, if there's less need for endoscopy, many of these ancillary services will contract.

Today, the opportunity cost for gastroenterologists to move away from screening colonoscopy is still too high. My goal is to help my practice plan for the future by not waiting until the opportunity cost drops significantly before exploring other areas. We need to anticipate that this opportunity cost will decrease and start developing new approaches now.

Praveen: You've mentioned in the past that gastroenterologists must own the GI tract and develop programs that service the entire GI tract, moving away from solely focusing on colonoscopy. Can you elaborate on this?

Dr. Ketover: Yes, I started thinking about this years ago when I began doing capsule endoscopy with Given Imaging's PillCam. The original commercial name was M2A, which stood for "mouth to anus." They quickly realized that the name wasn't consumer-friendly, so they changed it. However, it made me consider that the GI tract starts at the tongue. When we swallow something, it's within the GI tract's domain. We should see our role as treating everything from swallowing a bolus to the exit of unused portions of what we've swallowed, considering the entire GI tract as our domain.

The opportunity cost to move resources away from procedures is still too high for most practices to invest. In our practice, we've tried to create centers of excellence around non-colonoscopy issues like advanced esophageal disease, inflammatory bowel disease, liver disease, functional and motility disorders, and celiac disease. These focus on treating more of the patient than just the endoscopic portion. We need to interface better with ENT, pulmonary medicine, urology, and colorectal surgery, especially for pelvic floor issues. However, the delay in this movement is primarily due to the high reimbursement rates for endoscopy procedures, making it difficult to shift resources away from them.

Praveen: How do you think that shift will happen, Scott? There's so much dependence on reimbursement, and it doesn't seem to be going away quickly. It's steadily declining, so there isn't a strong drive to take immediate action. When do you think the shift to focusing on other aspects of GI care, which comes with an opportunity cost, will happen?

Dr. Ketover: The counterargument is that, on a population basis, there are too few gastroenterologists in the nation. We need more gastroenterologists than we have today, which should give us more leverage. There's more cancer screening to be done, more endoscopic treatments, and more infusible drugs available. So, you would think we would be in a strong position regarding the leverage of our clinical skills. However, we're competing with large organizations that have billions of dollars in assets. While they talk about quality and outcomes, they're primarily focused on the cost of care. How do we deliver reasonably high-quality care nationwide at a lower cost?

To get gastroenterologists to that point, we need to find a new way to practice our skills and patient care that separates it from the fee-for-service, piecemeal revenue model. Personally, I believe the key to this shift is data. Data is king. The explosion of electronic and digital data is also happening in medicine, but clinicians haven't fully felt its impact yet.

There is an opportunity for gastroenterologists, whether they're in MSOs or private equity-backed practices, to integrate around data. Having a large common database controlled by gastroenterologists and their entities can help answer clinical questions and prove where cost-effectiveness and clinical benefits are truly present.

Praveen: Do you have any digital health initiatives going on at MNGI?

Dr. Ketover: Yes, we are deeply involved in telehealth and have been using our EHR for two decades. However, I believe that the major advancements in database management of clinical issues will come from external sources rather than EHR vendors, who are not moving quickly enough. Third-party collaborations will be crucial. If gastroenterologists don't pool their resources to address this, the void will be filled by large tech companies, and we will become just another cog in their wheel.

We are generating the data and providing patient care, but currently, that data is locked within our EHRs. We can't effectively mine it to help patients and advance our field. We need to find a way to manage this data ourselves. At MNGI, we are actively exploring opportunities and talking to potential collaborators. However, this effort can't be limited to just MNGI. It must involve MNGI and numerous other GI practices, whether they are with private equity or not, including independent practices and hospital systems. We need to create a unified database around GI conditions to lead this change.

Praveen: On one side, you have your GI practice, but on the other side, there are several startups helping the same patients through digital means, such as an app with an artificial intelligence algorithm. There might be clinicians involved, but they are creating a one-to-many solution. Have you thought about how this competition will evolve for you, or is it not a concern at this stage?

Dr. Ketover: It is a concern. You're right; with artificial intelligence and databases of effective clinical practices, patients will seek these solutions. Long before Western medicine accepted things like acupuncture, patients were already using it. They sought acupuncture, hypnosis therapy, various kinds of massage therapy, and other treatments that we once considered fringe, but patients found relief.

Your question highlights that entities can develop electronic means to bring help directly into the patient's hands via their smartphone, bypassing the physician altogether. How will GI practice cope with that? It will likely mean we focus more on disease entities requiring a direct relationship with the patient, driven by evidence-based medicine.

Praveen: What does the future of GI look like from this point on?

Dr. Ketover: In the short term, the outlook is quite good. There is still a significant reason for physicians' practices and hospitals to invest in endoscopic units and procedures. We are on the cusp of seeing effective therapeutics for diseases that previously had no treatment options. We're now looking at treating NASH and fibrotic liver disease with new drugs, whether they are infusible or oral. As GI practice continues to advance, I believe there will remain a strong emphasis on procedures, and that's appropriate.

However, we also need to step back and consider how to improve our cognitive work. We should devote time to developing programs that enhance patients' lives beyond just treating them endoscopically. From a surgical perspective, these are challenging questions, but I believe the future of GI remains bright. As I said, if it goes in your mouth, it's within our domain, and we should embrace that and seek ways to keep people healthy.

Praveen: With that future in mind, what actions must gastroenterologists take today? What foundation must be laid to create a future where GI care means much more than endoscopies, covering everything from the mouth to the anus?

Dr. Ketover: We're still in a siloed world. My practice, your practice, this hospital, this system - there are still many silos in the delivery of GI care. Traditionally, we've been concerned about and afraid to share our data with other silos because it might weaken us or strengthen them. We need to move beyond that mindset. We need to look at these individual silos and figure out how to create a network that improves all of them.

I'm really focused on data. I believe there will soon be opportunities to network practices, whether they're independent, employed, or backed by private equity, to collect, aggregate, analyze, and clinically use data in a way that benefits everyone. This includes all practices, all systems, and, most importantly, the patients.

Praveen: Thank you very much. Is there anything else you wanted to share that I didn't ask?

Dr. Ketover: No, not specifically. But you're doing a phenomenal job with the follow-up to your book and all these interviews. You're keeping us informed, and I learn much more than I contribute. You've helped me be a taker as much as a giver. So, thank you.

Praveen: I'm so glad to hear that. Thank you so much for saying so.

Update from Dr. Ketover (June 2024):

MNGI Digestive Health now has 105 gastroenterologists, 1,200 staff, and 33 endoscopy suites.

Payor updates are not keeping pace with expense inflation. The margin on ASC endoscopy remains good but is declining. Efficiency in endoscopy suite utilization is critical. Unfortunately, most payors do not support their "talk" of delivering high-quality services with the "walk" of higher payments for quality. Practices may need to turn to hospitals to provide "call" pay for delivering excellent 24/7/365 hospital coverage.

MOVING ON FROM GERM THEORY TO PRECISION MEDICINE: PROF. DAVID WHITCOMB OF ARIEL PRECISION MEDICINE

Professor David Whitcomb is the cofounder and chairman of Ariel Precision Medicine. He's widely considered the father of pancreatic genetics.

Themes: *chronic pancreatitis, personalized healthcare, precision medicine, genetics, predictive models, AI in medicine, clinical decision support, disease prevention*

Date: *November 2023*

Three Takeaways:

1. **Revolutionizing Diagnosis and Treatment with Precision Medicine:** Precision medicine aims to provide the right medicine, at the right dose, to the right patient, for the right disease, at the right time. This approach not only aims to improve patient outcomes but also to reduce healthcare costs by minimizing ineffective treatments and diagnostic odysseys. Ariel Precision Medicine exemplifies this by integrating genetic insights with clinical decision support, offering a model for the future of personalized healthcare.
2. **Challenges and Opportunities in Healthcare Innovation:** Despite the clear benefits of precision medicine, there are significant barriers to its adoption, including resistance from established healthcare systems and financial disincentives for change. However, the push for precision medicine represents a significant opportunity to empower patients and physicians with better tools for diagnosis and treatment, potentially transforming

the healthcare landscape and setting a new standard for patient care.

3. **The Future of Medicine - Integrative and Preventative Approaches:** The future of gastroenterology, and medicine in general, lies in a more integrative approach that combines genetics, epigenetics, and advanced data analytics to understand and treat diseases at their root cause. This paradigm shift from treating symptoms to preventing diseases through early diagnosis and personalized treatment plans promises to enhance the quality of care, reduce unnecessary treatments, and ultimately improve patient outcomes on a large scale.

Praveen: Let's start with the very basics. What is precision medicine?

Prof. David Whitcomb: Precision medicine is a concept that addresses a significant shortcoming in modern Western medicine, which often adopts a one-size-fits-all approach. The idea behind precision medicine is to provide the right medicine in the right dose to the right patient for the right disease at the right time, aiming for better outcomes and cost savings. This concept acknowledges that every person is different, primarily due to their genetic makeup, which is influenced by age, environmental factors, and habits.

Traditional Mendelian genetics, which suggests that one genetic factor causes a complex disease, applies to only about one to two percent of patients with a syndrome—a disease defined by a combination of signs and symptoms without specifying the underlying cause or treatment. Precision medicine seeks to personalize care by using early diagnosis to determine what the disease is and what it is not, employing biomarkers to predict disease severity, choosing effective medications, and ensuring these medications will help rather than harm the patient.

While everyone agrees on the need for precision medicine, the challenge lies in implementing this revolutionary idea. The traditional medical paradigm, based on the germ theory of disease, posits that one factor causes a complex disorder, identifiable by comparing populations with and without the disease. However, this approach falls short when dealing with the complexity of the human body and genome, which comprises over a billion base pairs with millions of variations. We need a new paradigm that can handle this complexity. Precision medicine, as I see it, requires this new approach to be effectively put into action.

Praveen: Going back to the history of medicine, when did we come up with the whole concept of precision medicine? How many years has it been?

Prof. Whitcomb: The concept of precision medicine began emerging in the early 80s, particularly in the field of cancer research, where it became clear that both an initial injury and a subsequent driver are necessary for cancer to develop. This was a departure from the simpler germ theory of disease, where one factor was thought to cause a complex disorder.

Historically, medicine has recognized variability in disease presentation. For instance, some individuals may carry an infectious agent without displaying symptoms, while others exhibit disease symptoms without detectable agents. This variability hinted at the need for a more personalized approach.

In the 1800s, medicine in the United States was chaotic, with numerous competing forms of medicine, including traditional, holistic, and even snake oil remedies. To bring order, a group of physicians in New York City decided that medicine should be based on science, education, and ethics. Significant progress was made in the 19th century, notably with the invention of the compound microscope, which allowed for the visualization of cells and germs.

Despite these advancements, medical education was unstructured until the early 1900s. A political solution was sought, leading to the Flexner Report, funded by the Carnegie Foundation. Abraham Flexner's report advocated for scientific rigor and educational standards in medicine, resulting in state licensing requirements for medical practice.

Over a century later, medical education remains largely influenced by the Flexner Report, which adheres to principles from the 1800s. This rigid curriculum has made it challenging to incorporate new ideas and concepts, including precision medicine. Physicians trained extensively in the germ theory of disease and evidence-based medicine find it difficult to transition to a paradigm that considers the complex interplay of genetics, environment, and individual variability.

This historical context explains why precision medicine, despite its potential, has struggled to gain widespread acceptance and implementation in modern medical practice.

Praveen: Amazing. If the Holy Grail of precision medicine is, let's say, a ten, where is the world right now? On a scale of one to ten?

Prof. Whitcomb: I would say it's at about two to three. There are some areas, like in cancer, where there's precision oncology, but that's really looking at the genome of the tumor, not the patient. There are some other areas where progress is being made. The concept is now more accepted, but people, especially physicians, don't know how to apply it. They often dislike genetics, and insurance companies don't cover it because their advising physicians deem it worthless. So, it's been a real problem.

Praveen: Very interesting. David, please share a little bit more about Ariel Precision Medicine. What does the company do?

Prof. Whitcomb: When I finished my training at Duke University and came to the University of Pittsburgh, I got a position at the VA. I wanted to crack the code on pancreatitis, especially chronic pancreatitis, because I had seen a couple of patients at Duke with this condition. My mentor said they were alcoholics. I insisted these old ladies weren't alcoholics—they were conservative Baptists who didn't drink and didn't lie. They had advanced chronic pancreatitis and had been seeking help everywhere. It had to be something other than alcohol. My mentor said it was a hopeless condition because the organ is destroyed before you can figure out what's going on with it. He warned me not to throw my career away and to go into hepatology instead.

However, I decided to focus on pancreatitis. When I got to Pittsburgh and studied 100 consecutive patients, I found they were all different. I reasoned it had to be something genetic. So, I contacted a couple of friends, and we started the Midwest Multicenter Pancreatic Study Group to work together. I chose hereditary pancreatitis, and Dr. Larry Gates, a friend in Lexington, Kentucky, found a family. We did the genetics and discovered it was indeed genetic. This key discovery unlocked many problems associated with the disease. We found that even people with the worst gene had a huge spectrum of the disease—some never got it, some got it as babies, and some developed fibrosis, diabetes, or cancer. This allowed us to start cracking the code of chronic pancreatitis.

Praveen: What prompted you to start Ariel Precision Medicine?

Prof. Whitcomb: I started a study called the North American Pancreatitis Study, getting 30 friends to help deeply phenotype patients with pancreatitis in a mechanistic way. It looked like an epidemiology study with extra questions, but it was really mechanistic. We discovered what the average person does, who has a genetic cause, and how different factors work together. However, we couldn't implement it because insurance companies wouldn't pay for it, and other companies offering it couldn't provide proper interpretations. Hospital informatics systems were restrictive, and even as division chief, I faced many barriers. Physicians were especially resistant to genetics due to its complexity.

We started Ariel Precision Medicine to provide clinical decision support because doctors don't just want to know the genetics—they want to know what to do. They need specific, actionable information with the rationale behind it. Ariel integrates all the information and gives physicians precise recommendations. Initially, we did deep sequencing, but reimbursement was an issue.

In January of this year, we flipped the script at Ariel. Instead of deep dives, we took known genetic variants, placed them on an SNP array with 800,000 other variants, and grouped them together. We do genotyping on individuals with more than 800,000 Single Nucleotide Polymorphisms (SNPs), putting it in the cloud. Once technically validated, we provide clinically interpreted reports to patients quickly.

We found pharmacogenetic genes on there, too, allowing us to predict drug efficacy and toxicity. We even have a phone app for immediate drug recommendations based on a patient's genetic profile. However, we can't legally tell doctors what to do due to malpractice risks.

To address this, I started a publishing company called Systems Modeling and Advanced Report Technology for Medical Decisions (Smart-MD) with a journal called the Smart-MD Journal of Precision Medicine. This allows us to provide expert-driven clinical decisions and resources to doctors, streamlining their understanding of genetic information and its implications for patient care.

Praveen: Very impressive. How accurate is the output that comes from your screening tool?

Prof. Whitcomb: First of all, it's a screening tool designed to screen through things we already know. We also recognize that there's a lot we don't know. There are rare genetic variants not on the array that could be causing the disease, and we won't detect them. However, it at least gives the doctor a direction and a priority. This type of genotyping is about 97 to 99 percent accurate. It's not 100 percent accurate, so if something is critical, you have to do DNA sequencing to verify it.

Regarding artificial intelligence, we don't use it for several reasons. The FDA has specific criteria for AI because you often don't know how the computer came up with an association or recommendation. It might sound good and may be possible, but is it true? If you had different patients from different populations, would you get a different answer? We don't know. Ariel uses mathematical modeling, where every step is explicit, and we can demonstrate exactly how we arrived at an answer supported by peer-reviewed literature. This ensures traceability and transparency, although it means we lag about three to five years behind the latest knowledge.

AI is incredibly powerful with structured data, such as in digital pathology, radiology, and endoscopy. It can replicate expert opinions in diagnosing diseases. However, there are concerns. AI often relies on data and training from the past, reflecting outdated concepts. It also struggles with parsing complex mechanisms in medical records, defined by syndromes and billing codes rather than mechanisms. This can lead to garbage in, garbage out (GIGO) situations. Lastly, AI can produce hallucinations—answers that sound plausible but aren't true. These issues make it difficult to retro-engineer and verify AI-generated conclusions.

Praveen: David, I would have agreed completely in the pre-generative AI days, especially with structured data and all that. In fact, it performs better with unstructured data than a human-built model. It's been shockingly fascinating. Again, this is not even a question; I'm prompted to just react to what you just said.

Prof. Whitcomb: The problem is not unstructured data; it's bad data and false data. And so we have to overcome that.

This is why I'm working on this SMART-MD Journal of Precision Medicine (smart-md.org) because I'm demanding that the authors are able to explain what they mean, and we put it in a way that's unequivocal. That way, when you pull something, you're pulling the right thing because it's tagged correctly. When AI does figure out something, the answers have to conform to the rules of physics and mechanisms, or they are unreliable. Before we can accept it as true, we have to be able to take the AI-generated insights and reduce them to things that we could have or should have recognized but didn't. Once we look at it, sort out the underlying mechanism, and determine that it is actually true, then we can adopt it. That's why we use AI all the time for hypothesis generation but not for reporting possible hallucinations to patients.

Praveen: Are patients ordering the test or clinicians?

Prof. Whitcomb: This is a medical test, so it has to be ordered by a physician, who is responsible for the interpretation, explanation, and change in treatment strategy. So they're the ones that order it. The problem is that currently, the insurance doesn't pay for it, so the patient has to pay for it out of pocket. But what they do get is a copy of the report with tools on their smartphone to access it, with very simple information they can share with any doctor in an emergency room or their pharmacist, that kind of thing. So that's the way it's set up.

Praveen: Does the patient pay you or pay the physician?

Prof. Whitcomb: They pay us.

Praveen: They pay you directly. But does the test have to be prescribed by a physician?

Prof. Whitcomb: Yes. By a physician, a pharmacist, or a PA—somebody in the United States with an NPI number, which is a National Provider Number. Other systems may work in other countries.

Praveen: How much does the test cost?

Prof. Whitcomb: For pharmacogenetics, which includes just about all the genes that the FDA recommends, it was introduced at about $250, but the cost of everything is going up, so it will be a little more in the future. For the pancreatic cancer screening test, which we're super excited about because it fills a huge gap, who do you actually pull the fire alarm on because they are a high-risk individual (HRI) and start searching for a hidden cancer? That test was introduced at $299. The pancreatitis test is also $299. However, once you get one test, a major portion of the cost of genotyping is already covered, so we give about a $100 discount on additional tests.

Praveen: In your work with Ariel Precision Medicine, especially when incorporating concepts from epigenetics and the microbiome, have you discovered new insights that have led to a different understanding?

Prof. Whitcomb: Absolutely. The beauty of the pancreas is that there are three cell types, and each does one thing. We know how it works; we know the mechanisms, which genes are expressed, and what they do. Additionally, with pancreatitis, you know when the process starts—it starts with an episode of acute pancreatitis. You can follow the trajectory because you have a starting point and measures of what's going wrong and how fast.

This has helped us understand not just the acinar, duct, and islet cells but also variations in the inflammatory process, such as who has exaggerated inflammation and who has rapid fibrosis, which is related to the immune system. We also gained tremendous insights into pain. For instance, looking at a patient with pancreatitis and a CT scan, no one can tell how much pain they have because the pain process depends on how the nervous system responds to injury, not on the pancreatic cells themselves.

Through genetic analysis, we found that people with severe pain often had high levels of genetics for depression and anxiety, amplifying their perception of pain. We also discovered genes associated with chronic low back pain, shoulder pain, post-mastectomy pain, and injury-related pain, indicating their bodies can't easily tamp down the pain. This is related to the brain's descending inhibition, which can be verified through quantitative sensory testing.

We also understand maldigestion and overlapping inflammatory diseases by looking at pancreatic damage, giving us insight into how the inflammatory system works. This has helped us crack the code for more complex diseases like celiac disease, inflammatory bowel disease, and liver disease. We now have a handle on the keys to the wrong immune and pain responses.

We're developing modules for fatty liver disease progression, diarrhea with abdominal pain, and other conditions. We're also gaining insights into cardiovascular disease, dyslipidemia, cholesterol, and different types of diabetes—whether it's type 1, type 2, or type 3, insulin resistance, beta cell failure, susceptibility to inflammation, and the effectiveness of medications like metformin. All of this will be coming out soon from us.

Praveen: Professor David Whitcomb, it's been a pleasure talking to you. This has been such an insightful conversation. I'm deeply impressed with your mission, and I wish you and your team all the best.

Prof. Whitcomb: Thank you for having me. I appreciate what you're doing because there are a lot of innovations and people working hard. It's hard to get the message out. Your work, your platform, and your vision for what the future can be are very important.

WE NEED MORE HEALTH PROFESSIONALS ON SOCIAL MEDIA: DR. AUSTIN CHIANG OF MEDTRONIC ENDOSCOPY

Dr. Austin Chiang is a gastroenterologist, chief medical officer for Medtronic's Endoscopy arm, and a prominent social media influencer pioneering healthcare education and outreach.

Themes: *social media influence, digital health, interdisciplinary approach, integrating diverse perspectives, MedTech, combating misinformation*

Date: *December 2023*

Three Takeaways:

1. **The Power of Social Media in Healthcare**: Dr. Austin Chiang emphasizes the significant role of social media in raising awareness and educating the public about healthcare. His journey from using platforms like Twitter and TikTok to becoming a respected voice in medicine showcases the potential of social media to reach a wide audience, combat misinformation, and even influence corporate decisions and public health policies.
2. **Embracing Innovation and Technology**: Dr. Chiang's involvement in MedTech and his forward-looking view on the future of gastroenterology highlight the importance of embracing innovation. He discusses the advancements in AI, MedTech, and digital health, which are transforming how procedures are performed, improving efficiency, and ultimately enhancing patient care.
3. **Interdisciplinary Approach and Continuous Learning**: Dr. Chiang's career reflects the value of an interdisciplinary approach and continuous learning. His experiences

in clinical practice, MedTech, social media, and academic research demonstrate the importance of integrating diverse perspectives to drive progress.

Praveen: Austin, you're an interventional endoscopist, CMO of Medtronic Endoscopy, and a big social media influencer making a difference to many, many lives. First, let's go back all the way. Why did you decide to become a doctor?

Dr. Austin Chiang: There are several reasons. There are several physicians in my family. My grandfather was a surgeon in the World War II era. I have several cousins who are also physicians, none of whom are gastroenterologists. However, that was my environment and upbringing. My father is an engineer. He worked for a medical device company previously at one point at Baxter. So, I feel like, in the back of my mind, there was this environment that I grew up in. On top of that, my parents exposed me to a lot of great experiences growing up, like volunteering, sometimes in medical settings. Ultimately, when I was in college, I tried to keep an open mind with a lot of different classes, even many that were not relevant to my major, which was biology, but also political science, economics, etc. And I tried to be honest with myself about what would be a good fit for my personality. Ultimately, I felt that medicine was the right path, so that was my entry into the whole space.

Praveen: How did you pick GI?

Dr. Chiang: Again, I kept a very open mind going into medical school and went through all the rotations thinking that I originally was going to be a surgeon. So, I was definitely interested in something procedural. Along the way, I thought that I might be interested in plastic surgery, orthopedic surgery, and neurology at one point, and it was really the internal medicine residents that inspired me to go into internal medicine. Originally, I thought I would pursue pulmonary critical care. However, once I entered residency, I changed my mind, thought I was going to do interventional cardiology, and, at the last minute, decided to shift to gastroenterology. Part of that was because of my friends, honestly, who were all applying to gastroenterology, and I wondered why all my friends were going to GI. That told me I would mesh well with the personality in gastroenterology. But on top of that, I discovered that there's a large variety of organs that the field offers a lot of technology and innovation, taking things that were more invasive surgically to something less invasive, which is kind of a lot of what procedural GI is based upon. And that technology aspect has guided me toward where I've ended up today.

Praveen: If you don't mind, Austin, I want you to brag a little about the impact your social media work has had. You've innovated in social media as well. How do you do it?

Dr. Chiang: On the surface, it looks like it doesn't take much effort, and maybe some platforms, especially if you're focused on one thing, it is not much effort. But when you think about all the different platforms out there and creating content across the board, it can be a lot. But this journey really goes a while back. I've always been a lover of social media. I even used social media before Facebook times, like when Myspace and other blogging sites were in vogue. Facebook came out my freshman year in college, so I was the target demographic then. Over the years, I've seen how social media can drive awareness and captivate audiences outside of healthcare. I believed in its reach. Ultimately, when I was in residency a little over a decade ago, I spent some time at ABC News trying to learn how people receive medical information through conventional media because I noticed how some of my patients were coming to the hospital after things that they'd heard on TV. So, I was curious to know what that process was like.

I spent some time behind the scenes at ABC News as part of the medical unit, and at that time, they hosted weekly Twitter chats about various health topics. That's when I thought, *We could use more trained health professionals speaking out on these platforms, speaking about what we're trained to talk about rather than leaving it to people who aren't trained and have a lot of misinformation.*

From there, I started out on Twitter. I started noticing that some patients were paying attention, and this was just me talking about what I was learning in training or what my experience was at various conferences. Twitter is kind of, the bar is pretty low for entry, and it's all kind of text-based, especially back then. Slowly, as I completed training, I felt more comfortable getting on Instagram, putting my face out there, and talking about health on that platform.

Similarly, I've always just been an early adopter of all the various platforms. So, on TikTok in 2019, I was probably one of very few doctors on TikTok at the time. I actually went back and told all of my Instagram friends, most of whom were not in gastroenterology, actually probably all of whom were not in gastroenterology. The pandemic really then accelerated the conversation further because everyone was on their phones, everyone was on social media, spending a lot of time, and it really highlighted the importance of how social media can impact public opinion of healthcare and public perception of health professionals. The number one health topic, or the number one topic on the news for a long time, was the pandemic. So it gave a lot of us doctors a lot of visibility.

However, I will say that back in 2019, I founded a nonprofit professional society called the Association for Healthcare Social Media to help educate other health professionals on how to use social media. We partnered with various platforms like YouTube, LinkedIn, and Pinterest to educate on how not only to use the platforms but also how to create content, paying attention to lighting and audio and things that impact the quality of the production. We also partnered with Cochrane Collaborations to give out many healthcare social media research grants. We treated it like an academic endeavor.

Throughout the years, I was encouraged by some of my mentors to treat this whole social media presence like an academic pursuit. I actually did a lot of social media-related research on that, which was published in various GI journals and at conferences. What a lot of people don't realize is that this is not just me posting on social media. There was a bigger undercurrent, where I've learned a lot and gained a lot of traction in many ways.

To this day, I've taken that experience and applied it to my academic position, where at that institution, I was once chief medical social media officer, helping shape the social media policy and helping out with strategy. In my role now as chief medical officer, I'm seeing how we can be more patient-facing and still help YouTube with its misinformation efforts in collaboration with some other organizations like the American Board of Internal Medicine. It goes far beyond what people often see in the world of GI, and it's given me a lot of great opportunities. The latest of which was being involved in the healthcare leaders' social media roundtable for the White House. That was a big deal.

Praveen: That's awesome. What's been the impact in numbers and beyond?

Dr. Chiang: Directly measuring impact is tough, but my main goal was to reach patients and the general public more than my physician colleagues. Empowering patients has always been my focus. Some platforms have been better than others for this. In terms of metrics, my combined follower/subscriber count is over 650,000, and on TikTok alone, my videos have over 180 million views. Some specific videos have been very impactful. For example, a video on bariatric endoscopy generated multiple patient referrals to major academic centers around the country. Another instance is when a startup showed my TikTok to Medtronic's CEO, Geoff Martha, which helped them secure an investment. That TikTok was one of the first times Geoff came across me before meeting me in person, highlighting the power of social media even in a corporate setting.

Praveen: Very interesting. Austin, what did you learn from all this?

Dr. Chiang: The biggest lesson I've learned is the importance of sticking to an idea I truly believe in and finding champions to support me. Social media is a tool that should be used in healthcare, just as it impacts every other sector. Even in 2023, we haven't fully harnessed its potential to raise awareness and educate. Another key lesson is the importance of storytelling and communication. This didn't come naturally to me; I honed it through practice, especially by creating short-form content. The challenge was to distill complex concepts into engaging, 15-second pieces that people could easily understand. This skill has improved my general communication, helping me convey messages more effectively to colleagues and patients.

Medical education focuses heavily on science and medicine, which is crucial, but we need more training in communication, the business of healthcare, health policy, and technology. These areas are essential for better serving our patients and communities, and I'm hoping to contribute to that change.

Praveen: I've met you in person. You come through as quiet, serious, even reticent. How does such a personality transform on the platforms?

Dr. Chiang: Honestly, I am a bit of an introvert at baseline and very professional in the workplace. Anyone who works with me at the hospital knows I'm quiet and serious. It's funny because some patients find me on TikTok and come to see me expecting a totally different person.

At first, it was uncomfortable. I didn't expect anyone I knew to see my TikTok, but as more people did, I thought, *The cat's out of the bag,* so I just went with it. People still take me seriously because I've always focused on building a strong clinical foundation and aimed to train at the best places with respected mentors.

Everyone has a fun side, even the most serious people in gastroenterology. They might not publicly show it, but they have it. I'm just more out there and comfortable presenting it to the outside world now. It's about striking a balance. I speak differently on TikTok than on LinkedIn, where my audience is mainly corporate professionals.

Praveen: My final question, Austin, is your view on the future of gastroenterology. Where do you see things going three years from now, five years from now, even ten years from now?

Dr. Chiang: Gastroenterology is such a broad field. While we at Medtronic are focused on MedTech, there's a lot happening in pharmaceuticals with conditions like the microbiome and obesity. We'll continue to learn more over the years. Specifically for MedTech, we're just seeing the beginning of AI, which will improve clinical outcomes and operational efficiency. This could help address physician burnout by improving workflow and efficiency.

There are still other improvements to be made in the device space to make our lives easier and more ergonomic. Since my training, we've seen many surgical procedures shift to gastroenterology, moving toward less invasive methods, and that trend will continue.

Lastly, putting on my social media hat, I believe we'll see a more informed general public. That's my goal, especially with my book and social media presence. Hopefully, this will lead to people getting the appropriate care they need and better utilizing healthcare resources. I hope to see and contribute to progress in that area.

Praveen: Superb. On that note, Austin, what would be your final piece of advice to people?

Dr. Chiang: I would say always try to consider a different perspective. What has led me to where I am today is always trying to understand not necessarily what I believe in but why people who disagree with me think the way they do. Be open to unconventional ideas. In my current role in MedTech, I see not only the clinical practice side but also the challenges in how innovation is developed and brought to market. This understanding helps me see the pain points in the MedTech business that ultimately impact how we perform our procedures and affect patients' lives. So, if things aren't going your way or don't align with your beliefs, question it and dig deeper. That's how I approach everything.

Praveen: That piece of advice is golden. I thoroughly enjoyed our conversation. Thanks once again.

Dr. Chiang: Thank you. It's been an honor, Praveen. Thank you so much for having me.

ACKNOWLEDGMENTS

The Shift wouldn't have been possible without the countless conversations at the *GI Mastermind* program and during *The Scope Forward Show*. You are the pioneers of this field - thank you for all that you do.

Thank you to my clients. Your questions have prompted me on this journey to discover the future in a different way.

My team at NextServices – you are amazing. Your courage to experiment and pave the way for the future teaches me so much. Thank you for giving me the space and time necessary for an endeavor like this book.

Peter Diamandis and the team at Singularity University and Abundance 360, in more ways than one, have seeded ideas since 2012 and helped me connect the dots to the future of gastroenterology. Thank you.

Being the odd one out in a family is never easy. Thank you so much for always letting me do my own thing.

I am deeply grateful to the incredible individuals who generously shared their time and expertise for this book. My heartfelt thanks go to Dunston Almeida, Dr. Russ Arjal, Dr. Erica Barnell, Dr. Michael Byrne, Dr. Sanket Chauhan, Dr. Austin Chiang, Dr. Fehmida Chipty, Dr. Andrea Cherubini, Dr. Sonia Grego, Sam Holliday, Saurabh Jejurikar, Sam Jactel, Dr. Scott Ketover, Dr. Lawrence Kosinski, Asaf Kraus, Dr. Bara El Kurdi, Dr. Aja McCutchen, Dr. Dan Neumann, Dr. Jonathan Ng, Dr. Greg O'Grady, Dr. Michael Owens, Dr. Treta Purohit, Dr. Megan Riehl, Matt Schwartz, Dr. Madison Simons, Aonghus Shortt, Torrey Smith, Dr. Stephen Steinberg, Dr. Brennan Spiegel, Erik Duhaime, Omer Dror, and Professor David Whitcomb.

Thank you, Niloufer Venkataraman, former editor-in-chief of *National Geographic Traveller India* and *National Geographic Magazine*. It was fantastic to collaborate with you again. Your mastery of your craft is a blessing for writers. Thank you, Igniting Souls for stepping in during the final stage of this project and bringing this book to life. A special shoutout to Dr. Treta Purohit for reading the manuscript and providing invaluable feedback, especially for the clinical narrative for 2034.

ChatGPT – you are a darling. It's been a journey of both pain and enlightenment having you as a constant companion since your birth in November 2022. Working with you is both humbling and worrisome.

Finally, gratitude to the teachers of the Himalayan Tradition. The practices have expanded my mind in ways that can't easily be described, helping to bring this book to life.

NOTES

FOREWORD

Wikipedia, "Second Derivative," Last Accessed July 24, 2024, https://en.wikipedia.org/wiki/Second_derivative.

"Economic Recovery Act of 2009." CMS.gov. Last Accessed July 24, 2024. https://www.cms.gov/medicare/regulations-guidance/legislation/economic-recovery-act-2009.

Wikipedia, "Affordable Care Act," Last Accessed July 24, 2024, https://en.wikipedia.org/wiki/Affordable_Care_Act.

Wikipedia, "Center for Medicare and Medicaid Innovation," Last Accessed July 24, 2024, https://en.wikipedia.org/wiki/Center_for_Medicare_and_Medicaid_Innovation.

"'What is CMMI?' and 11 other FAQs about the CMS Innovation Center." KFF, February 27, 2018. https://www.kff.org/medicare/fact-sheet/what-is-cmmi-and-11-other-faqs-about-the-cms-innovation-center/.

"MACRA." CMS.gov. Last Accessed July 24, 2024. https://www.cms.gov/medicare/quality/value-based-programs/chip-reauthorization-act.

"RBRVS overview." American Medical Association. Last Accessed July 24, 2024. https://www.ama-assn.org/about/rvs-update-committee-ruc/rbrvs-overview.

Dr. Kosinski, L., email message to author, August 4, 2024.

INTRODUCTION

"Tiny Health." Last Accessed July 24, 2024. https://www.tinyhealth.com/store/baby-gut-health-test.

Dwoskin, E., Gilbert, D., and Hunter T. "Doctors couldn't help. They turned to a shadow system of DIY medical tests." The Washington Post, June 9, 2024. https://www.washingtonpost.com/technology/2024/06/09/home-health-tests-doctors-fda/.

Wible, P. "When doctors commit suicide, it's often hushed up." The Washington Post, July 14, 2014. https://www.washingtonpost.com/national/health-science/when-doctors-commit-suicide-its-often-hushed-up/2014/07/14/d8f6eda8-e0fb-11e3-9743-bb9b59cde7b9_story.html.

"The future of healthcare: The end of digital health." Rock Health, October 5, 2020. https://rockhealth.com/insights/the-future-of-healthcare-the-end-of-digital-health/.

Yahoo Finance. "NASDAQ Composite Index (^IXIC) Chart." Last Accessed July 24, 2024. https://finance.yahoo.com/chart/%5EIXIC.

Dr. Murali, N.S., email message to author, July 16, 2014.

Andersen, H. The Emperor's New Clothes. New York: George H. Doran Company, 1923.

EmTech Digital 2022.

Farr, C. "Proteus Digital Health, once valued at $1.5 billion, files for Chapter 11 bankruptcy." CNBC, June 16, 2020. https://www.cnbc.com/2020/06/15/proteus-digital-health-once-worth-1point5-billion-files-for-chapter-11.html.

"NextServices." Last Accessed July 24, 2024. https://nextservices.com/.

Wikipedia, "Giant sucking sound," Last Accessed July 24, 2024, https://en.wikipedia.org/wiki/Giant_sucking_sound.

Suthrum, P. Scope Forward: The Future of Gastroenterology Is in Your Hands. Mumbai: Lioncrest Publishing, 2020.

Suthrum, P. Private Equity in Gastroenterology: Navigating the Next Wave. Self-published, 2019.

"GI Mastermind." Last Accessed July 24, 2024. https://nextservices.com/gi-mastermind-2024.

"Scope Forward." Last Accessed July 24, 2024. https://scopeforward.com/.

CHAPTER 1

Wikipedia, "Khardung La," Last Accessed July 24, 2024, https://en.wikipedia.org/wiki/Khardung_La.

Duggan, W. "A Short History Of The Great Recession." Forbes, June 21, 2023. https://www.forbes.com/advisor/investing/great-recession/.

Wikipedia, "Great Depression," Last Accessed July 24, 2024, https://en.wikipedia.org/wiki/Great_Depression.

Friedman T.L. Thank You for Being Late: An Optimist's Guide to Thriving in the Age of Accelerations. New York: Farrar, Straus and Giroux, 2016.

Sorkin, A.R., and Peters, J.W. "Google to Acquire YouTube for $1.65 Billion." The New York Times, October 9, 2006. https://www.nytimes.com/2006/10/09/business/09cnd-deal.html.

Markoff, J. "Apple Introduces Innovative Cellphone." The New York Times, January 10, 2007. https://www.nytimes.com/2007/01/10/technology/10apple.html.

Wikipedia, "Spotify," Last Accessed July 24, 2024, https://en.wikipedia.org/wiki/Spotify.

Metz, R. "Twitter takes e-chat to extremes." NBC News, April 13, 2007. https://www.nbcnews.com/id/wbna18078020.

"Industry Leaders Announce Open Platform for Mobile Devices." Open Handset Alliance, November 5, 2007. https://www.openhandsetalliance.com/press_110507.html.

"Introducing Amazon Kindle." Amazon.com, November 19, 2007. https://press.aboutamazon.com/2007/11/introducing-amazon-kindle.

Schonfeld, E. "AirBed And Breakfast Takes Pad Crashing To A Whole New Level." TechCrunch, August 11, 2008. https://techcrunch.com/2008/08/11/airbed-and-breakfast-takes-pad-crashing-to-a-whole-new-level/.

Markoff, J. "Computer Wins on 'Jeopardy!': Trivial, It's Not." The New York Times, February 16, 2011. https://www.nytimes.com/2011/02/17/science/17jeopardy-watson.html.

Helft, M. "Netflix to Deliver Movies to the PC." The New York Times, January 16, 2007. https://www.nytimes.com/2007/01/16/technology/16netflix.html.

National Human Genome Research Institute. "The Cost of Sequencing a Human Genome." Fact Sheets. Accessed June 26, 2024. https://www.genome.gov/about-genomics/fact-sheets/Sequencing-Human-Genome-cost.

National Institutes of Health. "NIH Human Microbiome Project Defines Normal Bacterial Makeup of the Body." NIH News Releases, June 13, 2012. https://www.nih.gov/news-events/news-releases/nih-human-microbiome-project-defines-normal-bacterial-makeup-body.

Nakamoto, S. "Bitcoin: A Peer-to-Peer Electronic Cash System." Bitcoin.org. Last Accessed July 24, 2024. https://bitcoin.org/bitcoin.pdf.

"Microsoft Cloud Services Vision Becomes Reality With Launch of Windows Azure Platform." Microsoft, November 17, 2009. https://news.microsoft.com/2009/11/17/microsoft-cloud-services-vision-becomes-reality-with-launch-of-windows-azure-platform/.

Olson, P. "Facebook Closes $19 Billion WhatsApp Deal." Forbes, October 6, 2014. https://www.forbes.com/sites/parmyolson/2014/10/06/facebook-closes-19-billion-whatsapp-deal/.

Blystone, D. "The History of Uber." Investopedia, February 14, 2024. https://www.investopedia.com/articles/personal-finance/111015/story-uber.asp#toc-an-idea-is-born.

Grady, D. "Two Tests Added to Recommended List to Prevent or Detect Colorectal Cancer." The New York Times, March 6, 2008. https://www.nytimes.com/2008/03/06/health/research/06cancer.html.

Ahlquist, D. A., Sargent, D. J., Loprinzi, C. L., Levin, T. R., Rex, D. K., Ahnen, D. J., Knigge, K., Lance, M. P., Burgart, L. J., Hamilton, S. R., Allison, J. E., Lawson, M. J., Devens, M. E., Harrington, J. J., and Hillman, S. L. "Stool DNA and occult blood testing for screen detection of colorectal neoplasia." Annals of Internal Medicine, October 7, 2008. https://doi.org/10.7326/0003-4819-149-7-200810070-00004.

"Exact Sciences Extends, Expands Relationship with Mayo Clinic." Exact Sciences Corp, May 17, 2012. https://investor.exactsciences.com/investor-relations/press-releases/press-release-details/2012/Exact-Sciences-Extends-Expands-Relationship-with-Mayo-Clinic/default.aspx.

"FDA Approves Exact Sciences' Cologuard®; First and Only Stool DNA Noninvasive Colorectal Cancer Screening Test." Exact Sciences Corp, August 12, 2014. https://investor.exactsciences.com/investor-relations/press-releases/press-release-details/2014/FDA-Approves-Exact-Sciences-Cologuard-First-and-Only-Stool-DNA-Noninvasive-Colorectal-Cancer-Screening-Test/default.aspx.

"EXACT SCIENCES ANNOUNCES FINAL NATIONAL COVERAGE DETERMINATION FOR COLOGUARD®." Exact Sciences Corp, October 9, 2014. https://www.exactsciences.com/newsroom/press-releases/exact-sciences-announces-final-national-coverage-determination-for-cologuard.

Exact Sciences Corporation. "2023 Investor Day Presentation." PDF Document. Last Accessed July 24, 2024. http://q4live.s22.clientfiles.s3-website-us-east-1.amazonaws.com/877809405/files/doc_events/2023/Jun/21/exas-2023-investor-day-presentation.pdf.

"NEXT-GENERATION COLOGUARD® TEST DEMONSTRATES HIGH SENSITIVITY AND SPECIFICITY IN PIVOTAL BLUE-C STUDY, SIGNIFICANTLY OUTPERFORMING FECAL IMMUNOCHEMICAL TESTING (FIT) FOR CANCER AND PRECANCER DETECTION."

Exact Sciences Corp, October 22, 2023. https://www.exactsciences.com/newsroom/press-releases/next-generation-cologuard-test-demonstrates-high-sensitivity-and-specificity-in-pivotal-blue-c-stud.

"The Platinum Standard." US Digestive Health. Accessed June 23, 2024. https://usdigestivehealth.com/your-digestive-health/procedures/colonoscopies/the-platinum-standard/.

Hochman, A. "Math Teachers Stage a Calculated Protest." The Washington Post, April 3, 1986. https://www.washingtonpost.com/archive/local/1986/04/04/math-teachers-stage-a-calculated-protest/c003ddaf-b86f-4f2b-92ca-08533f3a5896/.

Diamandis, P. H. "THE SIX DS OF EXPONENTIALS." Peter H. Diamandis, October 28, 2020. https://www.diamandis.com/blog/6-ds-exponentials.

Wikipedia, "COVID-19 pandemic," Last Accessed July 24, 2024, https://en.wikipedia.org/wiki/COVID-19_pandemic.

"Geneoscopy." Crunchbase. Last Accessed July 24, 2024. https://www.crunchbase.com/organization/geneoscopy.

"Geneoscopy Raises $6.9 Million to Advance Preventive Screening Test for Colorectal Cancer and Advanced Adenomas." Geneoscopy, Inc., October 2, 2019. https://www.geneoscopy.com/geneoscopy-raises-6-9-million-to-advance-preventive-screening-test-for-colorectal-cancer-and-advanced-adenomas/.

"Geneoscopy Closes $105M in Financing to Advance its Noninvasive Multifactor RNA Screening Test for Colorectal Cancer Prevention." PR Newswire, November 16, 2021. https://www.prnewswire.com/news-releases/geneoscopy-closes-105m-in-financing-to-advance-its-noninvasive-multifactor-rna-screening-test-for-colorectal-cancer-prevention-301424501.html.

"Iterative Health." Crunchbase. Last Accessed July 24, 2024. https://www.crunchbase.com/organization/iterative-health.

"Iterative Scopes Closes $5.2M Seed Fundraising Round." PR Newswire, January 15, 2020. https://www.prnewswire.com/news-releases/iterative-scopes-closes-5-2m-seed-fundraising-round-300987016.html.

"Iterative Scopes Raises $30 Million Series A Financing to Advance AI-Driven Precision Medicine for Gastroenterology." Business Wire, August 3, 2021. https://www.businesswire.com/news/home/20210803005407/en/Iterative-Scopes-Raises-30-Million-Series-A-Financing-to-Advance-AI-Driven-Precision-Medicine-for-Gastroenterology.

"Iterative Scopes Announces $150 Million Series B to Advance AI-Driven Precision Medicine for Gastroenterology." Business Wire, January 19, 2022. https://www.businesswire.com/news/home/20220119005231/en/Iterative-Scopes-Announces-150-Million-Series-B-to-Advance-AI-Driven-Precision-Medicine-for-Gastroenterology.

"Oshi Health." Crunchbase. Last Accessed July 24, 2024. https://www.crunchbase.com/organization/oshi-health-inc.

"Oshi Health secures $23 million Series A to scale transformational virtual specialty care." Oshi Health, Inc., October 12, 2021. https://oshihealth.com/oshi-health-secures-23-million-series-a-to-scale-transformational-virtual-specialty-care.

"Oshi Health Raises $30M Series B Funding to Scale Access to Its Virtual Multidisciplinary Digestive Care." Oshi Health, Inc., April 11, 2023. https://oshihealth.com/oshi-health-raises-30m-series-b-funding-to-scale-access-to-its-virtual-multidisciplinary-digestive-care/.

"Virgo Surgical Video Solutions." Crunchbase. Last Accessed July 24, 2024. https://www.crunchbase.com/organization/virgo-marketing.

"Olympus Innovation Ventures Backs Endoscopy Video and AI Company, Virgo Surgical Video Solutions." PR Newswire, July 27, 2022. https://prnewswire.com/news-releases/olympus-innovation-ventures-backs-endoscopy-video-and-ai-company-virgo-surgical-video-solutions-301594359.html.

Dr. Erica Barnell (Cofounder, CMO, and CSO, Geneoscopy), interview with author, December 2023.

"FoodMarble Launches Next Generation WellTech Digestive Tracking Device for Consumers." Health Tech Digital, February 11, 2022. https://www.healthtechdigital.com/foodmarble-launches-next-generation-welltech-digestive-tracking-device-for-consumers/.

"FoodMarble." Crunchbase. Last Accessed July 24, 2024. https://www.crunchbase.com/organization/foodmarble.

Winn, Z. "Gamifying medical data labeling to advance AI." MIT News Office, June 28, 2023. https://news.mit.edu/2023/gamifying-medical-data-labeling-ai-0628.

Ang, A. "Alimetry gets FDA clearance for wearable gut disorder test." MobiHealthNews, June 16, 2022. https://www.mobihealthnews.com/news/anz/alimetry-gets-fda-clearance-wearable-gut-disorder-test.

"Alimetry." Crunchbase. Last Accessed July 24, 2024. https://www.crunchbase.com/organization/alimetry.

"Dieta Health raises $1.2M to Personalize Digestive Healthcare." Dieta Health, August 19, 2021. https://dietahealth.com/blog/dieta-pre-seed-funding.

"Dieta Health." Crunchbase. Last Accessed July 24, 2024. https://www.crunchbase.com/organization/dieta.

Kelly, S. "Medtronic Rolls Out AI-Assisted Colonoscopy System in Europe." MedTech Dive, October 21, 2019. https://www.medtechdive.com/news/medtronic-rolls-out-ai-assisted-colonoscopy-system-in-europe/565431/.

"EndoSound is the Winner of the 2022 Shark Tank Competition at the AGA Center for GI Innovation and Technology." EndoSound, May 5, 2022. https://www.endosound.com/endosound-is-the-winner-of-the-2022-shark-tank-competition-at-the-aga-center-for-gi-innovation-and-technology/.

"EndoSound." Crunchbase. Last Accessed July 24, 2024. https://www.crunchbase.com/organization/endosound.

"H1 2024 digital health funding: Resilience leads to brilliance." Rock Health, July 8, 2024. https://rockhealth.com/insights/the-future-of-healthcare-the-end-of-digital-health/.

Hirsch, J. "Elon Musk: Model S not a car but a 'sophisticated computer on wheels.'" Los Angeles Times, March 19, 2015. https://www.latimes.com/business/autos/la-fi-hy-musk-computer-on-wheels-20150319- 14 story.html.

Wikipedia, "Moore's Law," Last Accessed July 24, 2024. https://en.wikipedia.org/wiki/Semiconductor.

Moore, G. E. "Cramming more components onto integrated circuits." Computer History Museum, April 19, 1965. https://www.computerhistory.org/collections/catalog/102770822.

"GRAPHCORE TAKES ON NVIDIA WITH LATEST AI CHIP." Graphcore. Last Accessed July 24, 2024. https://www.graphcore.ai/articles/graphcore-takes-on-nvidia-with-latest-ai-chip.

Hu, K. "ChatGPT sets record for fastest-growing user base - analyst note." Reuters, February 2, 2023. https://www.reuters.com/technology/chatgpt-sets-record-fastest-growing-user-base-analyst-note-2023-02-01/.

Howarth, J. "How Many People Own Smartphones? (2024-2029)." Exploding Topics, June 13, 2024. https://explodingtopics.com/blog/smartphone-stats.

Suthrum, P. Scope Forward: The Future of Gastroenterology Is in Your Hands. Mumbai: Lioncrest Publishing, 2020.

Friedman T.L. Thank You for Being Late: An Optimist's Guide to Thriving in the Age of Accelerations. New York: Farrar, Straus and Giroux, 2016.

Diamandis, P. H. "THE ONGOING MARCH OF MOORE'S LAW." Peter H. Diamandis, October 29, 2023. https://www.diamandis.com/blog/scaling-abundance-series-16.

Wikipedia, "Jack Welch," Last Accessed July 24, 2024. https://en.wikipedia.org/wiki/Jack_Welch.

Wikipedia, "Ernest Shackleton," Last Accessed July 24, 2024. https://en.wikipedia.org/wiki/Ernest_Shackleton.

"Carol Dweck: A Summary of Growth and Fixed Mindsets." Farnam Street. Last Accessed June 26, 2024. https://fs.blog/carol-dweck-mindset/.

Wikipedia, "Carol Dweck," Last Accessed July 24, 2024. https://en.wikipedia.org/wiki/Carol_Dweck.

CHAPTER 2

Sam Jactel (Founder and CEO, Ayble Health), interview with author, December 2022.

Dr. Jonathan Ng (CEO and Founder, Iterative Health), interview with author, October 2022.

Ismail, S., Palao, F., Lapierre, M. Exponential Transformation: Evolve Your Organization (and Change the World) With a 10-Week ExO Sprint. New Jersey: John Wiley & Sons, Inc., 2019.

Ismail, S., Malone, M.S., Geest, Y. Exponential Organizations: Why new organizations are ten times better, faster, and cheaper than yours (and what to do about it). New York: Diversion Books, 2014.

"99Designs." Last Accessed July 24, 2024. https://99designs.com/.

"PatientsLikeMe." Last Accessed July 24, 2024. https://www.patientslikeme.com/.

Asaf Kraus (Founder and CEO, Dieta Health), interview with author, September 2023.

Erik Duhaime (Founder and CEO, Centaur Labs), interview with author, August 2022.

Dr. Daniel Neumann (President and CSO, Capital Digestive Care | Cofounder, Trillium Health), interview with author, July 2023.

Dunston Almeida (Founder and CEO, triValence), interview with author, November 2023.

"TriValence." Last Accessed July 24, 2024. https://www.trivalence.com/.

"Thinkific." Last Accessed July 24, 2024. https://www.thinkific.com/.

"GI Mastermind." Last Accessed July 24, 2024. https://nextservices.com/gi-mastermind-2024.

"HeyGen." Last Accessed July 24, 2024. https://www.heygen.com/.

"Calendly." Last Accessed July 24, 2024. https://calendly.com/.

Anthony, S. D. "Kodak's Downfall Wasn't About Technology." Harvard Business Review, July 15, 2016. https://hbr.org/2016/07/kodaks-downfall-wasnt-about-technology.

"Freelancer." Last Accessed July 24, 2024. https://www.freelancer.com/.

"Fiverr." Last Accessed July 24, 2024. https://www.fiverr.com/.

"Toptal." Last Accessed July 24, 2024. https://www.toptal.com/.

"Plaeto." Last Accessed July 24, 2024. https://plaeto.in/.

"BlandAI." Last Accessed July 24, 2024. https://www.bland.ai/.

"Microsoft Power Automate." Last Accessed July 24, 2024. https://make.powerautomate.com/.

"Power BI." Last Accessed July 24, 2024. https://www.microsoft.com/en-us/power-platform/products/power-bi/.

"ChatGPT." Last Accessed July 24, 2024. https://openai.com/chatgpt/.

"Claude." Last Accessed July 24, 2024. https://claude.ai/.

CHAPTER 3

"Medscape Gastroenterologist Lifestyle, Happiness & Burnout Report 2023: 'Contentment Amid Stress'" Medscape, Accessed June 26, 2024. https://www.medscape.com/slideshow/2023-lifestyle-gastroenterologist-6016078#8.

"Medscape Physician Burnout & Depression Report 2024: 'We Have Much Work to Do.'" Medscape, Accessed June 26, 2024. https://www.medscape.com/slideshow/2024-lifestyle-burnout-6016865.

"Physician Compensation Report 2024." Doximity. Last Accessed June 27, 2024. https://www.doximity.com/reports/physician-compensation-report/2024.

Anderson, J. C., Bilal, M., Burke, C. A., Gaidos, J. K., Lopez, R., Oxentenko, A. S., Surawicz, C. M. "Burnout Among US Gastroenterologists and Fellows in Training: Identifying Contributing Factors and Offering Solutions." National Library of Medicine. Last Accessed June 26, 2024. https://pubmed.ncbi.nlm.nih.gov/36477385/.

Dr. Fehmida Chipty (Gastroenterologist, Founder, Zamana Art), interview with author, May 2022.

Press, E. "The Moral Crisis of America's Doctors." The New York Times, June 15, 2023. https://www.nytimes.com/2023/06/15/magazine/doctors-moral-crises.html.

"Ameca: The Future Face of Robotics." Engineered Arts. Last Accessed July 24, 2024. https://www.engineeredarts.co.uk/robot/ameca/.

Dr. Sanket Chauhan (Founder and CEO, Surgical Automations), interview with author, December 2023.

"Surgical Automations." Last Accessed July 24, 2024. https://autosurg.com/.

Harari, Y. N. Homo Deus: A Brief History of Tomorrow. New York: HarperCollins, 2017.

GZERO Staff. "Yuval Noah Harari: AI Is a Social Weapon of Mass Destruction to Humanity." GZERO Media, March 12, 2024. https://www.gzeromedia.com/gzero-world-clips/yuval-noah-harari-ai-is-a-social-weapon-of-mass-destruction-to-humanity.

Harari, Y. N. "Yuval Noah Harari on what the year 2050 has in store for humankind." WIRED, August 12, 2018. https://www.wired.com/story/yuval-noah-harari-extract-21-lessons-for-the-21st-century/.

Dr. Russ Arjal (Cofounder and CMO, WovenX Health), interview with author, August 2023.

"Phil Knight Quotes." Goodreads. Last Accessed July 24, 2024. https://www.goodreads.com/author/quotes/3319233.Phil_Knight.

"WovenX." Last Accessed July 24, 2024. https://www.wovenxhealth.com/.

Dr. Daniel Neumann (President and CSO, Capital Digestive Care, Cofounder, Trillium Health), interview with author, July 2023.

"Trillium Health." Last Accessed July 24, 2024. https://www.trilliumhealth.org/.

Dr. Bara El Kurdi (Gastroenterologist, Host of the GI StartUp podcast), interview with author, September 2023.

Dr. Aja McCutchen (Cofounder, OLVI Health, Board member of United Digestive), interview with author, April 2024.

"OLVI Health." Last Accessed July 24, 2024. https://olvihealth.com/.

Dr. Lawrence Kosinski (Founder and CMO, SonarMD, Founder, VOCNomics), interview with author, September 2022.

"SonarMD." Last Accessed July 24, 2024. https://sonarmd.com/.

"VOCnomics." Last Accessed July 24, 2024. https://www.vocnomics.com/.

"SonarMD Chief Medical Officer Develops eNose Technology to Promote Gut Health." SonarMD, October 7, 2022. https://sonarmd.com/sonarmd-chief-medical-officer-develops-enose-technology-to-promote-gut-health/.

Dr. Scott Ketover (President and CEO, MNGI Digestive Health), interview with author, April 2021.

"The future of healthcare: The end of digital health." Rock Health, October 5, 2020. https://rockhealth.com/insights/the-future-of-healthcare-the-end-of-digital-health/.

CHAPTER 4

Levin, S. "Netflix cofounder: 'Blockbuster laughed at us ... Now there's one left.'" The Guardian, September 14, 2019. https://www.theguardian.com/media/2019/sep/14/netflix-marc-randolph-founder-blockbuster.

Brooks, M. "Downward Trend in Medicare Payments for GI Services." Medscape, November 08, 2022. https://www.medscape.com/viewarticle/983731.

Abundance360 2024.

Dr. Lawrence Kosinski (Founder and CMO, SonarMD, Founder, VOCNomics), interview with author, September 2022.

Dr. Scott Ketover (President and CEO, MNGI Digestive Health), interview with author, April 2021.

"How Many Pictures? Photo Statistics That Will Blow You Away." Light Stalking. Last Accessed June 30, 2024. https://www.lightstalking.com/photo-statistics/.

"Satisfai Health." Last Accessed July 24, 2024. https://www.satisfai.health/.

Wikipedia, "Turing Test," Last Accessed July 24, 2024, https://en.wikipedia.org/wiki/Turing_test.

"Turing Test in Artificial Intelligence." GeeksforGeeks. Last Accessed July 24, 2024. https://www.geeksforgeeks.org/turing-test-artificial-intelligence/.

Kurzweil, R. The Age of Spiritual Machines. New York: Penguin Books, 1999.

PowerfulJRE. "Joe Rogan Experience #2117 - Ray Kurzweil." YouTube video, 4.43. March 12, 2024. https://www.youtube.com/watch?v=w4vrOUau2iY.

Morris, C. "Elon Musk predicts AI will be smarter than humans by next year." Fortune, April 9, 2024. https://fortune.com/2024/04/09/elon-musk-ai-smarter-than-humans-by-next-year/.

Peter H. Diamandis. "Elon Musk on AGI Safety, Superintelligence, and Neuralink (2024)." YouTube video,1.01. March 26, 2024. https://www.youtube.com/watch?v=akXMYvKjUxM.

"OpenAI." Last Accessed July 24, 2024. https://openai.com/.

Bajekal, N, and Perrigo, B. "CEO of the Year 2023: Sam Altman." TIME, December 6, 2023. https://time.com/6342827/ceo-of-the-year-2023-sam-altman/.

TIME. "Sam Altman on OpenAI, Future Risks and Rewards, and Artificial General Intelligence." YouTube video, 10.15. https://www.youtube.com/watch?v=e1cf58VWzt8.

Aonghus Shortt (Founder and CEO of FoodMarble), interview with author, May 2022.

Dr. Sanket Chauhan (Founder and CEO of Surgical Automations), interview with author, December 2023.

Dr. Erica Barnell (Cofounder, CMO, and CSO, Geneoscopy), interview with author, December 2023.

Kolata, G. "F.D.A. Approves Blood Test for Colon Cancer Detection." The New York Times, July 29, 2024. https://www.nytimes.com/2024/07/29/health/colon-cancer-blood-test-shield.html.

Dr. Brennan Spiegel (Director of Health Services Research, Cedars-Sinai Health System and Author of VRx: How Virtual Therapeutics Will Revolutionize Medicine), interview with author, March 2021.

Spiegel, B. VRx: How Virtual Therapeutics Will Revolutionize Medicine. New York: Basic Books, 2020.

Torrey Smith (Cofounder and CEO, Endiatx), interview with author, March 2024.

Dr. Jonathan Ng (CEO and Founder, Iterative Health), interview with author, October 2022.

Tolle, E. A New Earth: Awakening to Your Life's Purpose. New York: Penguin Books, 2006.

VOICES I - DIAGNOSTICS AND DEVICES

Aonghus Shortt (Founder and CEO of FoodMarble), interview with author, May 2022.

"Foodmarble." Last Accessed July 24, 2024. https://foodmarble.com/.

"FoodMarble." Tracxn. Last Accessed July 24, 2024. https://tracxn.com/d/companies/foodmarble/__MBFRB9s56uGwnK4O9NxhI7zbwllUlSbKjVKi-gRg6zQ.

Shortt, A., email message to author, July 5, 2024.

Prof. Greg O'Grady (CEO and Cofounder, Alimetry, Professor of Surgery, The University of Auckland), interview with author, August 2022.

"Alimetry." Last Accessed July 24, 2024. https://www.alimetry.com/.

"Alimetry Raises $16.3 Million To Launch Wearable Diagnostic Device For The Gut." Scoop, March 2, 2022. https://www.scoop.co.nz/stories/SC2203/S00007/alimetry-raises-163-million-to-launch-wearable-diagnostic-device-for-the-gut.htm.

Ang, A. "Alimetry gets FDA clearance for wearable gut disorder test." MobiHealthNews, June 16, 2022. https://www.mobihealthnews.com/news/anz/alimetry-gets-fda-clearance-wearable-gut-disorder-test.

Sonia Grego, PhD (Duke Professor, Cofounder, and CEO of Coprata), interview with author, June 2022.

"Coprata." Last Accessed July 24, 2024. https://www.coprata.com/.

Dr. Grego, S., email message to author, June 28, 2024.

"Oura ring." Last Accessed July 24, 2024. https://ouraring.com/.

"WHOOP." Last Accessed July 24, 2024. https://www.whoop.com/in/en/.

Dr. Erica Barnell (Cofounder, CMO, and CSO, Geneoscopy), interview with author, December 2023.

"Geneoscopy." Last Accessed July 24, 2024. https://www.geneoscopy.com/.

"JAMA Publishes Geneoscopy's Pivotal CRC-PREVENT Trial Results, Reporting Highest Sensitivity for Detecting Colorectal Cancer and Advanced Adenomas Among Available Noninvasive Screening Tests." Geneoscopy, October 23, 2023. https://www.geneoscopy.com/jama-publishes-geneoscopys-pivotal-crc-prevent-trial-results-reporting-highest-sensitivity-for-detecting-colorectal-cancer-and-advanced-adenomas-among-available-noninvasive-screening-tests/.

IPM staff. "Grail Launches Medicare Multi-Cancer Early Detection Study." Inside Precision Medicine, November 22, 2023. https://www.insideprecisionmedicine.com/topics/oncology/grail-launches-medicare-multi-cancer-early-detection-study/.

"GRAIL." Last Accessed July 24, 2024. https://grail.com/.

"Freenome." Last Accessed July 24, 2024. https://www.freenome.com/.

Wikipedia, "SEPT9," Last Accessed July 24, 2024, https://en.wikipedia.org/wiki/SEPT9.

"Geneoscopy." Crunchbase. Last Accessed July 24, 2024. https://www.crunchbase.com/organization/geneoscopy.

"Geneoscopy Raises $6.9 Million to Advance Preventive Screening Test for Colorectal Cancer and Advanced Adenomas." Geneoscopy, Inc., October 2, 2019. https://www.geneoscopy.com/geneoscopy-raises-6-9-million-to-advance-preventive-screening-test-for-colorectal-cancer-and-advanced-adenomas/.

"Geneoscopy Closes $105M in Financing to Advance its Noninvasive Multifactor RNA Screening Test for Colorectal Cancer Prevention." PR Newswire, November 16, 2021. https://www.prnewswire.com/news-releases/geneoscopy-closes-105m-in-financing-to-advance-its-noninvasive-multifactor-rna-screening-test-for-colorectal-cancer-prevention-301424501.html.

Dr. Stephen Steinberg (Founder and CMO, EndoSound), interview with author, February 2024.

"EndoSound." Last Accessed July 24, 2024. https://www.endosound.com/.

"Sonivate Medical." Last Accessed July 24, 2024. https://sonivate.com/.

"FDA Grants Breakthrough Device Designation for Innovative EndoSound Vision Ultrasound System." EndoSound, August 31, 2021. https://www.endosound.com/fda-grants-breakthrough-device-designation-for-innovative-endosound-vision-ultrasound-system/.

"EndoSound is the Winner of the 2022 Shark Tank Competition at the AGA Center for GI Innovation and Technology." EndoSound, May 5th, 2022. https://www.endosound.com/endosound-is-the-winner-of-the-2022-shark-tank-competition-at-the-aga-center-for-gi-innovation-and-technology/.

"EndoSound." Crunchbase. Last Accessed July 24, 2024. https://www.crunchbase.com/organization/endosound.

VOICES II - AI AND ROBOTICS

Dr. Jonathan Ng (CEO and Founder, Iterative Health), interview with author, October 2022.

"Iterative Health." Last Accessed July 24, 2024. https://iterative.health/.

"Iterative Scopes Announces $150 Million Series B to Advance AI-Driven Precision Medicine for Gastroenterology." Business Wire, January 19, 2022. https://www.businesswire.com/news/home/20220119005231/en/Iterative-Scopes-Announces-150-Million-Series-B-to-Advance-AI-Driven-Precision-Medicine-for-Gastroenterology.

Matt Schwartz (CEO and Founder, Virgo Surgical Video Solutions), interview with author, June 2022.

"Virgo." Last Accessed July 24, 2024. https://virgosvs.com/.

Wikipedia, "Douglas Rex," Last Accessed July 24, 2024, https://en.wikipedia.org/wiki/Douglas_Rex.

Bhageshpur, K. "Data Is The New Oil -- And That's A Good Thing." Forbes, November 15, 2019. https://www.forbes.com/sites/forbestechcouncil/2019/11/15/data-is-the-new-oil-and-thats-a-good-thing/.

"Virgo Surgical Video Solutions." Crunchbase. Last Accessed July 24, 2024. https://www.crunchbase.com/organization/virgo-marketing.

"Olympus Innovation Ventures Backs Endoscopy Video and AI Company, Virgo Surgical Video Solutions." PR Newswire, July 27, 2022. https://prnewswire.com/news-releases/olympus-innovation-ventures-backs-endoscopy-video-and-ai-company-virgo-surgical-video-solutions-301594359.html.

Dr. Michael Byrne (CEO and Director, Satisfai Health), interview with author, September 2023.

"Satisfai Health." Last Accessed July 24, 2024. https://www.satisfai.health/.

"Satisfai Health Purchases Assets of Docbot, Inc.." Satisfai Health, November 21, 2022. https://www.satisfai.health/newsreleases/sustainable-eztbs.

"Satisfai Health and the University of Oxford formally collaborate in the field of AI for the evaluation of Barrett's Oesophagus and Oesophageal Cancer." Satisfai Health. October 2, 2023. https://www.satisfai.health/newsreleases/satisfai-health-and-the-university-of-oxford-formally-collaborate-in-the-field-of-ai-for-the-evaluation-of-barretts-oesophagus-and-oesophageal-cancer.

"Alimentiv, Satisfai Health, and Virgo Announce Partnership to use AI driven technology to Enhance Clinical Trials in IBD and other GI Diseases." Satisfai Health. October 5, 2023. https://www.satisfai.health/newsreleases/carbon-zero-a7mm7.

Byrne, M. F., Parsa, N., Alexandra T., Greenhill, Chahal, D., Ahmad, O., Bagci, U. AI in Clinical Medicine: A Practical Guide for Healthcare Professionals. New Jersey: John Wiley & Sons Ltd, 2023.

"Viz.ai." Last Accessed July 24, 2024. https://www.viz.ai/.

Wikipedia, "Moore's Law," Last Accessed July 24, 2024, https://en.wikipedia.org/wiki/Semiconductor.

Dr. Andrea Cherubini (SVP Science, AI and Data, Cosmo IMD), interview with author, March 2024.

"Cosmo IMD." Last Accessed July 24, 2024. https://cosmoimd.com/.

Kelly, S. "Medtronic Rolls Out AI-Assisted Colonoscopy System in Europe." MedTech Dive, October 21, 2019. https://www.medtechdive.com/news/medtronic-rolls-out-ai-assisted-colonoscopy-system-in-europe/565431/.

"U.S. FDA Grants De Novo Clearance for First and Only Artificial Intelligence System for Colonoscopy; Medtronic Launches GI Genius™ Intelligent Endoscopy Module." Medtronic, April 21, 2021. https://news.medtronic.com/2021-04-12-U-S-FDA-Grants-De-Novo-Clearance-for-First-and-Only-Artificial-Intelligence-System-for-Colonoscopy-Medtronic-Launches-GI-Genius-TM-Intelligent-Endoscopy-Module.

"Medtronic and NVIDIA Collaborate to Build AI Platform for Medical Devices." NVIDIA, March 21, 2023. https://investor.nvidia.com/news/press-release-details/2023/Medtronic-and-NVIDIA-Collaborate-to-Build-AI-Platform-for-Medical-Devices/.

Saurabh Jejurikar (Founder and CEO, EndoVisionAI), interview with author, November 2023.

"Endovision AI." Last Accessed July 24, 2024. https://www.endovisionai.com/.

Dr. Sanket Chauhan (Founder and CEO, Surgical Automations), interview with author, December 2023.

"Surgical Automations." Last Accessed July 24, 2024. https://autosurg.com/.

Goodman, B. "New study examines the effectiveness of colonoscopies." CNN, October 10, 2022. https://edition.cnn.com/2022/10/09/health/colonoscopy-cancer-death-study/index.html.

"Intuitive Surgical Operations, Inc." Last Accessed July 24, 2024. https://www.intuitive.com/en-in.

Wikipedia, "Frederic Moll," Last Accessed July 24, 2024, https://en.wikipedia.org/wiki/Frederic_Moll.

Torrey Smith (Cofounder and CEO, Endiatx), interview with author, March 2024.

"Endiatx." Last Accessed July 24, 2024. https://endiatx.com/.

"Endiatx." Crunchbase. Last Accessed July 24, 2024. https://www.crunchbase.com/organization/endiatx.

VOICES III - DATA AND OTHER INNOVATIONS

Omer Dror (Founder and CEO, Lynx.MD), interview with author, September 2023.

"Lynx.MD." Last Accessed July 24, 2024. https://lynx.md/.

Asaf Kraus (Founder and CEO, Dieta Health), interview with author, September 2023.

"Dieta Health." Last Accessed July 24, 2024. https://dietahealth.com/.

"Newly Published Mayo Clinic and VA Studies: Dieta's AI Enhances Management of Cirrhosis." Dieta Health, February 24, 2024. https://dietahealth.com/blog/mayovastudies.

"Dieta Health raises $1.2M to Personalize Digestive Healthcare." Dieta Health, August 19, 2021. https://dietahealth.com/blog/dieta-pre-seed-funding.

"Dieta Health." Crunchbase. Last Accessed July 24, 2024. https://www.crunchbase.com/organization/dieta.

Dunston Almeida (Founder and CEO, triValence), interview with author, November 2023.

"triValence." Last Accessed July 24, 2024. https://www.trivalence.com/.

Erik Duhaime (Founder and CEO, Centaur Labs), interview with author, August 2022.

"Centaur Labs." Last Accessed July 24, 2024. https://www.centaurlabs.com/.

"Eko Health, Inc.." Last Accessed July 24, 2024. https://www.ekohealth.com/.

"DiagnosUs." Last Accessed July 24, 2024. https://www.diagnosus.com/.

VOICES IV - NEW CARE MODELS

Sam Jactel (Founder and CEO, Ayble Health), interview with author, December 2022.

"AybleHealth." Last Accessed July 24, 2024. https://www.ayblehealth.com/.

Jactel, S., email message to author, July 10, 2024.

"Mahana." Last Accessed July 24, 2024. https://www.mahana.com/.

"metaMe Health." Last Accessed July 24, 2024. https://www.metamehealth.com/.

Nagel, T. "What Is It Like to Be a Bat?." Philosophical Review Duke University Press, Last Accessed July 15, 2024. https://www.sas.upenn.edu/~cavitch/pdf-library/Nagel_Bat.pdf.

Sam Holliday (CEO and Cofounder, Oshi Health) | Dr. Treta Purohit (Medical Director, Oshi Health), interview with author, April 2024.

"Oshi Health." Last Accessed July 24, 2024. https://oshihealth.com/.

"Oshi Health." Crunchbase. Last Accessed July 24, 2024. https://www.crunchbase.com/organization/oshi-health-inc.

"Oshi Health secures $23 million Series A to scale transformational virtual specialty care." Oshi Health, Inc., October 12, 2021. https://oshihealth.com/oshi-health-secures-23-million-series-a-to-scale-transformational-virtual-specialty-care.

"Oshi Health Raises $30M Series B Funding to Scale Access to Its Virtual Multidisciplinary Digestive Care." Oshi Health, Inc., April 11, 2023. https://oshihealth.com/oshi-health-raises-30m-series-b-funding-to-scale-access-to-its-virtual-multidisciplinary-digestive-care/.

"GI Mastermind." Last Accessed July 24, 2024. https://nextservices.com/gi-mastermind-2024.

Dr. Michael Owens (Physician and Cofounder, Pearl Health Partners), interview with author, October 2022.

"Pearl Healthcare Partners." Last Accessed July 24, 2024. https://pearlhealth.com/.

Dr. Megan Elizabeth Riehl (Clinical Assistant Professor, Michigan Medicine) and Dr. Madison Simons (GI Psychologist, Cleveland Clinic), interview with author, June 2022.

Dr. Brennan Spiegel (Director of Health Services Research, Cedars-Sinai Health System and Author of VRx: How Virtual Therapeutics Will Revolutionize Medicine), interview with author, March 2021.

Spiegel, B. VRx: How Virtual Therapeutics Will Revolutionize Medicine. New York: Basic Books, 2020.

VOICES V - ENTREPRENEURSHIP AND CAREERS

Dr. Fehmida Chipty (Boston-based gastroenterologist and Founder, Zamana Art), interview with author, May 2022.

Dr. Daniel Neumann (President and CSO, Capital Digestive Care | Cofounder, Trillium Health), interview with author, July 2023.

"Trillium Health." Last Accessed June 24, 2024. https://www.trilliumhealth.org/.

"GI Mastermind." Last Accessed July 24, 2024. https://nextservices.com/gi-mastermind-2024.

Dr. Aja McCutchen (Cofounder, OLVI Health and board member of United Digestive), interview with author, April 2024.

"OLVI Health." Last Accessed July 24, 2024. https://olvihealth.com/.

Dr. Bara El Kurdi (Gastroenterologist and Host of the GI StartUp podcast), interview with author, September 2023.

"GI StartUp Podcast." Last Accessed July 24, 2024. https://rss.com/podcasts/gistartup/.

Dr. Russ Arjal (Cofounder and CMO, WovenX Health), interview with author, August 2023.

Arjal, R., email message to author, July 15, 2024.

"WovenX." Last Accessed July 24, 2024. https://www.wovenxhealth.com/.

"Phil Knight Quotes." Goodreads. Last Accessed July 24, 2024. https://www.goodreads.com/author/quotes/3319233.Phil_Knight.

VOICES VI - PERSPECTIVES

Dr. Lawrence Kosinski (Founder and CMO, SonarMD and Founder, VOCnomics), interview with author, September 2022.

"SonarMD." Last Accessed July 24, 2024. https://sonarmd.com/.

"VOCnomics." Last Accessed July 24, 2024. https://www.vocnomics.com/.

"SonarMD Chief Medical Officer Develops eNose Technology to Promote Gut Health." SonarMD, October 7, 2022. https://sonarmd.com/sonarmd-chief-medical-officer-develops-enose-technology-to-promote-gut-health/.

Dr. Scott Ketover (President and CEO, MNGI Digestive Health), interview with author, April 2021.

"Surgical robots, early cancer tests top CB Insights digital health startup list." Medtech Dive, August 14, 2020. https://www.medtechdive.com/news/digital-health-companies-raise-20-billion-CB-insights/583489/.

Ketover, S., email message to author, June 18, 2024.

Prof. David C. Whitcomb (Founder and Chairman, Ariel Precision Medicine), interview with author, November 2023.

"Ariel Precision Medicine." Last Accessed July 24, 2024. https://arielmedicine.com/.

"Smart-md." Last Accessed July 24, 2024. https://smart-md.org/.

Wikipedia, "Flexner Report," Last Accessed July 24, 2024, https://en.wikipedia.org/wiki/Flexner_Report.

Dr. Austin Lee Chiang (CMO, Medtronic Endoscopy), interview with author, December 2024.

"Association for Healthcare Social Media." Last Accessed July 24, 2024. https://ahsm.org/.

Chiang, A. Gut: An Owner's Guide (The Body Literacy Library). London: Dorling Kindersley Limited, 2024

ABOUT THE AUTHOR

Praveen Suthrum is an entrepreneur and healthcare futurist. He cofounded NextServices, a company that helps healthcare organizations stay competitive by integrating AI, automation, and improving care delivery. Additionally, he runs a leadership program called the *GI Mastermind* and hosts a popular podcast called *The Scope Forward Show*.

Praveen has been featured in *Forbes, Inc., The Economic Times, The Detroit News, STAT News, Fortune,* and *Gastroenterology & Endoscopy News,* among other publications. He is the author of *Private Equity in Gastroenterology* and *Scope Forward* – books that have transformed the gastroenterology space. He has served as faculty at major gastroenterology societies, including the American College of Gastroenterology, the Digestive Health Physicians Association, and the American Society for Gastrointestinal Endoscopy.

He holds an engineering degree from Manipal Institute of Technology and an MBA from the University of Michigan's Ross School of Business. He also completed advanced yoga teacher training and yoga for cancer programs at The Yoga Institute.

Previously, he served as the Chairman of the Michigan Ross Alumni Board of Governors and on the board of Dr. Mohan's Diabetes Centre.

INSPIRED TO TAKE ACTION?

Praveen's initiatives in gastroenterology and healthcare:

NextServices
- Boost efficiency by automating your revenue cycle
- Build apps using a digital health studio
- Experience a refreshingly simple endo-writer without the bells and whistles

GI Mastermind
- A leadership program that gives you a front row seat to the future of gastroenterology

Scope Forward
- AI + human curated newsletter to stay relevant in a rapidly changing GI world

Explore next steps here:
scopeforward.com/transform

THIS BOOK IS PROTECTED INTELLECTUAL PROPERTY

The author of this book values Intellectual Property. The book you just read is protected by Instant IP™, a proprietary process, which integrates blockchain technology giving Intellectual Property "Global Protection." By creating a "Time-Stamped" smart contract that can never be tampered with or changed, we establish "First Use" that tracks back to the author.

Instant IP™ functions much like a Pre-Patent™ since it provides an immutable "First Use" of the Intellectual Property. This is achieved through our proprietary process of leveraging blockchain technology and smart contracts. As a result, proving "First Use" is simple through a global and verifiable smart contract. By protecting intellectual property with blockchain technology and smart contracts, we establish a "First to File" event.

Protected by Instant IP™

LEARN MORE AT INSTANTIP.TODAY

Printed in the USA
CPSIA information can be obtained
at www.ICGtesting.com
LVHW021304270924
792238LV00004B/877